PLANTS AND HABITATS

Exploratory Inspection Techniques

by

Gary R. Sanford, PhD

Copyright © 2024 Gary R. Sanford
All rights reserved.
ISBN: 9798326166869

Table of Contents

1: Introduction..1
    Suggested Tools..4
2: Existing Site Information...7
3: Types of Observations...13
    Sites of Interest..13
    Plant Observations..14
        Species Identification..15
        Plant Habits...15
        Functional Traits..16
    Plants in Space (Communities)................................17
4: Biotic Characteristics...21
    Plant Observations..21
        The Plant List..22
            Exercise #1: Plant List..............................23
        Growth Habit...23
            Exercise #2: Growth Habit.........................26
        Life Form..26
            Exercise #3: Life Form..............................28
        Life Duration...28
            Exercise #4: Life Duration..........................29
        Phenology..29
            Exercise #5: Phenology..............................31
        Plant Height...31
            Exercise #6: Plant Height............................34
        Tree Trunk Diameter...34
            Exercise #7: Tree Trunk Diameter................35
        Vegetative Reproduction....................................35
            Exercise #8: Vegetative Reproduction..........37
        Spinescence..38
            Exercise #9: Spinescence............................39
        Branching Architecture.....................................39

Exercise #10: Branching Intensity..............40
Leaf Size..40
   Exercise #11: Leaf Size............................43
Leaf Retention...44
   Exercise #12: Leaf Retention.....................45
Invertebrate Herbivore Damage...........................45
   Exercise #13: Invertebrate Leaf Damage.......46
Bark...46
   Exercise #14: Bark.................................49
Dispersal Syndrome................................50
   Exercise #15: Dispersal Syndrome..............51
Species Wetland Status...........................52
   Exercise #16: Wetland Indicator Category....54
Wetland Plant Morphology.....................54
   Exercise #17: Wetland Plant Morphology.....55
Salt Resistance.......................................55
   Exercise #18: Salt Resistance....................56
Community Structural Attributes.........................56
   Strata Characteristics.............................58
      Layering, Abundance, and Coverage.............58
         Exercise #19: Strata........................59
         Exercise #20: Abundance...................59
         Exercise #21: Species Coverage..............59
         Exercise #22: Layer Coverage..............62
      Gaps and Shading..............................62
         Exercise #23: Canopy Openness..............63
         Exercise #24: Canopy Gaps..................63
      Plant Distributions Within a Community.................64
         Exercise #25: Dispersion....................65
5: Abiotic Characteristics..67
   Landscape and Landform Features.....................67
      Location..68
         Exercise #26: Site Location................68
      Insolation......................................68

- Exercise #27: Shading............................68
- Exposure..69
  - Wind..69
    - Exercise #28: Wind Exposure......................70
  - Salt...70
    - Exercise #29: Salt Exposure........................71
- Landforms and Microfeatures................................71
  - Aspect..72
    - Exercise #30: Aspect..............................73
  - Flats..73
    - Exercise #31: Flats...............................74
  - Inclines..74
    - Exercise #32: Inclines............................78
  - Depressions.......................................78
    - Exercise #33: Depressions.......................78
- Geology and Soils...79
  - Geology..79
    - Bedrock..79
      - Exercise #34: Bedrock........................81
    - Unconsolidated Land Features.................81
      - Exercise #35: Unconsolidated Land Features 83
  - Soils..84
    - Soil Profiles.....................................91
      - Exercise #36: Soil Profile, Texture, and Color 91
    - Web Soil Survey..............................92
      - Exercise #37: Web Soil Survey Data...........98
    - Soil pH, Nutrition, and Other Soil Parameters. .98
      - Exercise #38: Soil Parameters From The Web 103
- Flooding and High Groundwater..............................104
  - Evidence of Temporary Inundation......................105
    - Exercise #39: Indicators of Past Inundation 106

Riparian Environments..107
    Exercise #40: Riparian Zone Identification 108
Wetland Hydrology...108
    Exercise #41: Water Sources and Flows.....112
    Exercise #42: Wetland Class........................112
Climate..112
    Exercise #43: Precipitation..........................113
    Exercise #44: Temperature..........................114
    Exercise #45: Growing Degree Days..........114
Appendix 1: Soil Drainage Classes...............................115
References..117

Table of Figures

Figure 1: Tree Height Measurements...............................32
Figure 2: Branching Intensity of *Ilex glabra*...................40
Figure 3: Examples of Leaf Shapes..................................43
Figure 4: Nested plot designs...58
Figure 5: Slope Complexity..76
Figure 6: Slope Shape...77
Figure 7: USDA Soil Texture Triangle.............................87
Figure 8: Texture by Feel Procedure.................................88
Figure 9: Soil pH, Ranges for pH classes, and Associated Soil Conditions..100

Index of Tables

Table 1: Site information and sources...............................10
Table 2: Commonly used plot sizes..................................20
Table 3: Growth Habits...23
Table 4: USDA Growth Habits and Definitions...............24
Table 5: Dispersal syndromes...50

Table 6: Morphological adaptations or responses to waterlogging..55
Table 7: Values for a coverage estimation technique.......60
Table 8: Physiognomic categories....................................62
Table 9: Dispersion Classes..65
Table 10: Definitions of Slope Classes............................74
Table 11: Some Slope and Angle Equivalencies..............74
Table 12: Sample Report from the Web Soil Survey........93
Table 13: Useful soil parameters....................................102

1: Introduction

When traveling the countryside, whether on foot or by vehicle, changes in vegetation can be noted that seem to correlate with various landscape features. For example, the types of plant communities growing along streams, lakes, and in wet depressions differ markedly compared to drier locations. Obvious differences exist between alpine communities and forested areas of lower elevation. On a larger scale, flying across the country allows for observations of forested and wooded landscapes, grasslands, and deserts that occur in various climate zones. Many of these correlations appear to have straight-forward explanations, although on closer inspection, controlling factors are often complex and identifying specific cause-and-effect relationships is difficult. Nevertheless, the process of understanding patterns of vegetation within a landscape begins with a close inspection of plants in the field.

Site inspections can be an important first step in correlating plant presence, abundance, condition, and distribution to habitat characteristics. Inspections promote the recognition of plant groupings that tend to recur across the landscape. It also allows the identification of areas where species composition changes along various types of gradients.

Inspections vary in their level of detail and rigor. They range from being casual and cursory in nature to reconnaissance-level outings used to identify areas for future research to making time-consuming quantitative field observations. For the serious field biologist, a casual and cursory inspection fails to provide more than an aesthetic appreciation of nature. Those doing research will understand the need for reconnaissance-level outings and subsequent detailed data acquisition. For many, an approach that exceeds casual observations but is less intensive than more research-oriented methods is needed.

Site inspection techniques deserve attention by students of

plant ecology, plant geography, field biology, and others interested in natural history. An inspection-level excursion is treated within this book as a field trip that can be accomplished within a few hours to a day by walking through sites of interest. It is intended to provide information that allows an appreciation of plants and vegetation at specific locations in context with attributes that may influence growth and development.

A plant may exist in a natural area when a group of co-existing organisms and an aggregate of physical and chemical conditions allow it to live and reproduce. Additional environmental features may exist in the area that have little or no influence on the plant. This entire complex of biotic and abiotic elements is referred to as the plant's habitat. Habitat can also refer to an area where conditions allow a group of populations, i.e., a community, to be found (e.g., bog, desert). Noting the presence and condition of a plant or community in its habitat, as well as the characteristics of the habitat itself, allows us to compare these various elements to those found at other locations. In doing so, it may be possible to speculate about what elements control the presence and abundance of individuals that comprise the local populations. Continued observations may lead to an understanding of the distribution patterns exhibited by a species and by a community within a landscape. Such observations can lead to correlations and hypotheses as to the driving forces behind these patterns.

Differences between locations relative to plant species composition, pattern of plant occurrence, abundance, or other biological characteristics may correlate with notable variations in habitat conditions. The correlations may be stochastic, but they become less likely to be a result of chance if they repeat across the landscape. Even with repeated observations, the complexity of a condition may confound the correlation. For example, a wetland obviously differs from the adjacent upland because of water. It also differs in many other ways, including soil (reducing conditions, oxygen availability, pH, nutrients, organic matter, toxic materials, etc.) and microclimate. So it is wrong to assume that ecological differences between upland and wetland sites are driven by water alone; the controlling agent may consist of the interwoven conditions of a wetland. Thus, while a habitat is composed of a complex of biotic and abiotic factors, a single factor may consist of a complex of constituents.

Habitat characteristics related to plant growth have been characterized and investigated in our modern era by numerous disciplines, including the sciences of plant ecology, plant physiology, forestry, grassland ecology, plant geography, and soil science. They have found a pivotal role in describing ecosystems, inventorying, describing, and monitoring land resources, generating ecological site descriptions, and conducting ecological integrity assessments of uplands, wetlands, and riparian environments (examples include: Luttmerding et al. 1990; British Columbia Ministry of Forests and Range and B.C.M.E. 2010; Herrick et al. 2017). Techniques included in this document have been drawn from various disciplines based on their ability to be incorporated into an inspection-level outing.

This book focuses on rapid inspection techniques done with minimal aid from instrumentation, which can be achieved during a walking tour. As a result, certain habitats are not considered since their examination requires specialized equipment, such as a boat, or they are relatively inaccessible. For example, river bottoms, deep lake habitats, and marine benthic zones are beyond the scope of the observations discussed here.

The second chapter of this book identifies some sources of information available prior to a field trip. Next, a chapter on units of observation is presented. Units range from plant parts to whole plants, groups of plants, and portions of landforms. Chapter 4 deals with specific observations and techniques useful for assessing the ecology of plants and communities. The last chapter considers abiotic components worthy of note during an inspection. An effort has been made to provide a synopsis of background information and rationales for the included techniques. Forty-five exercises are included that will aid in developing a record of observations.

Suggested Tools

One goal of this book is to offer evaluation techniques that do not require sophisticated instruments. Nevertheless, smart phones are ubiquitous, and their advantages may be put to good use with very little effort. These devices, with appropriate and readily available applications, can provide site-specific information on location, elevation, compass bearing, and distance between points. These data can also be determined from maps, such as USGS geodetic survey maps.

Latitude and longitude data can be shown on smart phones. The location of the phone at the time of observation is generally portrayed on a phone map. Various types of maps are available, including street and topographic maps and satellite imagery. Some phones and software will also list the device's elevation.

GPS accuracy depends on a number of properties, including the user's phone and software, the number of satellites accessible to the device, meteorological conditions, and interfering structures (see GPS.gov – How accurate is GPS?). Proper conditions may not be available at the time or location of observation.

Another useful instrument is the compass, some of which incorporate a clinometer and can be used to determine slope or plant heights. A pocket magnifying glass is encouraged. A pocket knife may occasionally be an advantage. If soils are of particular interest, then a short-handled drain spade, along with Munsell soil color charts, are suggested.

A pocket cloth tape measure can be useful when looking at

certain plant characteristics, such as the size of tree trunks. Choose one that has both English and metric scales. Metric scales are preferable for scientific purposes, however the English scale is more familiar to many in the United States. Other gear includes a waterproof notebook, a pencil, an indelible ink pen, and collecting bags. The notebook will serve for recording field observations. If permanent specimens are to be taken, then a plant press is recommended. Finally, knowing your pace length can allow pacing to help determine a variety of plant and terrain features.

Since an inspection is aimed at developing an approximate understanding of the vegetation, choosing a "typical" or representative plant for evaluation may be sufficient. With practice, a simple ocular estimate may be adequate for some parameters.

Students of plant ecology, field biology, and natural history can use the included techniques and observations to improve the usefulness of outings and field trips. The number and types of observations described herein may be too great to all be used during a single outing. Other types of observations beyond those covered in this book may be required in particular situations. The observer should be free to record and use whatever data fit the circumstances.

2: Existing Site Information

A survey of existing information concerning the locale one plans to visit is an appropriate first step. The depth of such a survey need not be great for an initial excursion, but the more prepared one is for a visit, the more fruitful the outing may be.

With the aid of the internet, an almost overwhelming amount of information can be obtained for just about any area in the US. Table 1 offers a variety of web links that may prove useful for an excursion.

United States Geological Survey maps are readily available and provide a wealth of information beyond just topography. Perennial and intermittent streams, wetland areas, woodland and shrubland areas, land use, and roads of all types are included. A perusal of map symbols (https://pubs.usgs.gov/gip/TopographicMapSymbols/topomapsymbols.pdf) provides a list of features that may be found.

Satellite imagery is readily available throughout the United States. Aerial orthophotographs of many areas can also be accessed. These are especially good as base maps, and if winter photos are available, evergreen and deciduous plant communities become obvious. Forest, shrubland, grassland, meadow, and marsh can be identified. Land use can be seen, as can edge environments. Old burns or other disturbances may become obvious. False-color infrared satellite images can be used to identify a number of features, such as the presence and health of vegetation. Aerial and satellite images can also be used to investigate historical conditions. See Riebeek (2014) for a quick guide to the interpretation of satellite imagery and Evans and Brewer (2015) for detailed information on vegetation image interpretation.

If an outing is planned along a recognized trail system, it is usually possible to obtain trail maps. With a little work, trail routes can be transferred to base imagery. GIS software (such as QGIS, which is free) can be used to combine information into one plan. While GIS software is complex and requires a long learning curve, other more easily mastered software packages, such as GIMP (the GNU Image Manipulation Program, also free), can be used to produce a base map with various types of added information. It is useful to place waypoints at recognizable locations so that one may occasionally orient oneself when in the field. Of course, it is possible to use a smart phone to locate your position and mark locations of interest on the phone's map as they are encountered. But the use of a paper map containing various types of information can be quite useful. Handwritten notes can be placed directly on the paper map, although a field notebook or data sheet should be used for recording observations discussed in Chapters 4 and 5. Because of its larger size, a paper map provides a broader view of the landscape compared to a cell phone map at comparable scales, allowing local areas to be put in perspective with surrounding habitats.

The US Fish & Wildlife Service National Wetlands Inventory mapper is useful in obtaining information on the locations and types of wetlands that may be present. The extent and type of wetlands can be shown on a variety of basemaps (ex., satellite photography, street maps, USGS topography maps).

Wetland types are based on the Federal Geographic Data Committee (2013) classification system. This classification uses a four-tier hierarchy, i.e., system, subsystem, class, and subclass. Five systems exist: marine, estuarine, riverine, lacustrine, and palustrine. Discussions of these systems are provided by the Federal Geographic Data Committee (2013).

Aquatic beds constitute a class found in all systems. These beds include rooted vascular plant communities in the marine system and both rooted and floating vascular plant communities found in all other systems. Additionally, emergent, scrub-shrub, and forested communities are classes found in the palustrine system. Emergent communities are classified into three subclasses: persistent, non-persistent, and *Phragmites australis*. Both scrub-shrub and forested communities include broad-leaved deciduous, needle-leaved deciduous, broad-leaved evergreen, needle-leaved evergreen, deciduous, and evergreen communities.

The United States Geological Service offers a large amount of data online. Bedrock and surficial geology maps are of particular interest for an outing. A general bedrock map of the U.S. showing geologic units (ex., igneous; metamorphic; sedimentary) is available at **https://mrdata.usgs.gov/geology/state/**. Detailed lithology may be downloaded for each state in the form of shapefiles (for use with GIS software) and KMZ files (for use with "Google Earth"). Google Earth can be directly accessed on the web or downloaded onto computers.

Each of over 26,000 geologic maps may be shown on screen using "MapView". As of this writing, the MapView beta version by the National Geologic Map Database is available online (see Table 1). Both bedrock and surficial maps are included in the collection.

The USDA Natural Resources Conservation Service is another source of valuable data. A map of the soils found in an area of interest can be produced on either a topographic or aerial photographic base map. Extensive information is available for each soil type. For example, the soil profile, drainage class, depth to the water table, frequency of flooding or ponding, and numerous chemical properties are listed.

Data related to the local climate are available from the National Weather Service. Look for maximum, minimum, and average temperature data, as well as precipitation and snowfall data. Average monthly data are also useful.

Maps, printed out in 8.5×11 inch sections and attached to a

clipboard, are quite useful for an initial field excursion. USGS maps can be used to determine slope orientation and incline. Aerial and satellite photographs provide information on vegetation, as do wetland maps. Soil maps show the location of various soil types. Placing all this information on one printout can obscure the plan, and it is recommended that several different maps be produced. Outlining wetland boundaries on orthophotographs or satellite photography allows identification of wetland areas while still leaving a clear view of vegetation. Separate topography and soil maps avoid confusing, overly detailed renditions. Place trails and/or observation points on all maps.

Table 1: Site information and sources

Type of Information	Source
Orthophotographs and Satellite Imagery (including color infrared Landsat)	Various government and private sites: (e.g., 1. Where to Get Data	Landsat Science (nasa.gov); 2. Get a map	Mass.gov; 3. World map, satellite view // Earth map online service (satellites.pro); 4. https://www.google.com/map/; 5. https://www.bing.com/maps/aerial?).
Aerial/Satellite Historical Images	1. https://earth.google.com/; 2. Where to Get Data	Landsat Science (nasa.gov).
Trail Maps	Various internet sources, including local, state, and federal governments.
USGS Topography	https://www.usgs.gov/programs/national-geospatial-program/topographic-maps.
Bedrock and Surficial Geology	1.https://mrdata.usgs.gov/geology/state/; 2. https://ngmdb.usgs.gov/mapview/?center=-97,39.6&zoom=4; 3. State government and university sources.

Type of Information	Source
Soils	https://websoilsurvey.sc.egov.usda.gov/App/HomePage.htm.
Wetlands and Riparian Habitats	https://www.fws.gov/program/national-wetlands-inventory/wetlands-mapper.
Weather Data	https://www.weather.gov/.

When visiting a new site, the number of plant species present can be overwhelming, yet at least a partial mastery of the flora is essential if correlations and hypotheses dealing with plant ecology are to be formulated. Good keys become essential for the identification of plants, and yet the sheer number of species turns the task of developing a species list for a site into an unmanageable endeavor on a first excursion. Developing a command of the flora will ultimately require an extensive period of time in a region. For example, over 1500 taxa are listed for the small area of Cape Cod in Massachusetts. Thousands of species may be listed in larger regional floras.

However, for a particular plant community, the list becomes more manageable. One or two hundred species per community is not unusual. Still, for an initial outing as envisioned in this book, this number is too large to rely on keys in the field. Knowledge of dominant and common species can make an excursion profitable. This may pare the list down to a few dozen plants, which makes field identification with a key practical.

It is of great help if data on flora and/or plant communities are available for a particular site. Local plant lists may be available on the web. Parks, particularly federal parks, may identify common plants and plant communities, and these lists may help on sites near parks. Sometimes, field guides are available for trails. Check trip guides and local government web sites for floristic information. Natural Heritage programs and state wildlife departments may offer plant community guides that identify dominant and indicator species for particular plant communities, as well as information on where communities may be found. Another source is the "Web-Soil Survey," which lists dominant plants for various soil types (click the Soil Data Explorer tab followed by the Ecological Sites tab).

By conducting a web search for species likely to be encountered on a trip, it will be possible to identify key traits and see photographs of each taxon. If an unknown species is encountered in the field and time is pressing, take photographs of the plant's habit, upper and lower leaf surfaces, flowers, and other potential key characteristics. Then assign a letter or number as an identification marker until the plant's species identity can be determined later. Obtaining a specimen for later perusal is, of course, ideal, if allowed by the landowner. If taking a specimen is allowable, then avoid taking rare plants.

3: Types of Observations

The numbers and types of observations that can provide ecologically meaningful ways to examine plants and communities are almost limitless. Consideration is restricted here to evaluations, estimates, and measurements that are appropriate for an inspection-level outing. A number of the evaluation techniques produce quantitative results. These results are intended to be used only to gain an approximate understanding of the vegetation since statistical accuracy cannot be determined by the procedures as described. The focus is to develop a general appreciation of plants and vegetation.

Observations related to plant parts, whole plants, species, and plant communities are often used in a comparative way. For example, comparing the vegetation found at several locations may lead to a correlation with certain habitat features. Changes in species composition and vegetation characteristics along an environmental gradient can lead to hypotheses concerning controlling factors.

Besides comparing two or more sites, it is also possible to compare plants or plant parts within one site. For example, comparing leaves found on the outer canopy and subject to high insolation with shaded leaves found deep within the canopy can elucidate phenotypic leaf plasticity.

Sites of Interest

"Sites of Interest" are areas of the landscape in which the techniques described in Chapters 4 and 5 are to be employed. It will be assumed here that an observer is interested in plants living in natural communities rather than anthropomorphically altered areas. For example, locations near a parking area or trail head often support weedy and invasive species. A trail itself may encourage the dispersal and colonization of these species, so off-trail observations are preferred. Therefore, try to find an area of interest away from altered habitat. This does not mean that naturally disturbed sites (e.g., by fire, windthrow, etc.) should be avoided. Also, successional sites or second-growth forests may be appropriate choices. Interest is centered on driving forces leading to natural communities.

Plant Observations

When examining a site, two types of observations are generally made: First is a consideration of individual plants. Plants grouped into a species represents an important unit for ecological investigations. But keep in mind that each individual has a group of identifiable parts (e.g., leaves, flowers, stems, roots, etc.), which can also be used to investigate ecological relationships. The second is how these plants are arranged in space. Spatial relationships occur in three dimensions, although a unit of land area may be used for such evaluations.

Making observations on individual plant characteristics requires selecting appropriate specimens. Selection guidelines are discussed in detail by Pérez-Harguindeguy et al. (2013). For purposes of site inspections where interest is not centered on a single taxon, selection criteria should emphasize common species found in each vegetation strata. Select around three representative individuals (ramets) per species for evaluation. Choose individuals that are reproductively mature and appear healthy. Exclude plants that have been heavily damaged by herbivores or pathogens (unless the goal is to assess such damage). If more than one characteristic is being evaluated, use the same individuals to assess as many characteristics as possible.

Species Identification

Plant features are used in order to assign an individual to its appropriate taxonomic group. Taxonomic groupings are designed to arrange species based on evolutionary relatedness. To achieve this goal, taxonomists utilize a variety of plant features, and historically, reproductive organs have been emphasized. These features are now augmented with genetic characteristics (e.g., chloroplast gene sequencing) that have led to revisions in taxonomic schemes. For the field ecologist, identifying a plant to the species level and beyond becomes a crucial activity that relies on observable traits with an emphasis on reproductive characteristics. Unfortunately, when visiting a site, many, if not most, plants will not be in reproductive mode, thus making identification more difficult.

When making a species identification, be cognizant of the type of habitat in which it grows. Many species are generalists that are able to survive in more than one habitat. For example, red maple (*Acer rubrum*) is found in both dry and wet sites. Other species are restricted to a particular type of habitat. Plant identification books and websites typically include information on species habitat affinity, and this information can help in identification.

Plant Habits

Other classifications exist that group individual plants. A plant's habit is an important feature that has ecological significance. A variety of habits are captured in the growth form designations used in the USDA Plants Data Base (see Table 4 in chapter 4). This system provides an immediate appreciation for the fundamental structure of a plant but lacks architectural specificity. For example, the system identifies trees and shrubs as classification units. However, a tree may be conical or broadly branching in overall shape. A shrub may arise at one point in the soil, while rhizomatous shrubs may produce interconnected networks of stems that are densely scattered across the ground surface.

Another method of ecologically grouping plants based on morphological characteristics (i.e., life forms) was developed by C. Raunkiaer and published in 1903 (see Raunkiaer, 1934). Though old, this form of grouping is still considered a simple and useful way of functionally classifying plants (Pérez-Harguindeguy et al., 2013). The Raunkiaer system is discussed under "Life Form" in chapter 4.

Functional Traits

Many morphological, anatomical, physiological, biochemical, and phenological characteristics of individual plants are important in describing the ecology of each plant, as well as its role in the plant community, higher trophic levels, and the ecosystem. These characteristics are referred to as functional traits. Some examples of functional traits include leaf size, stem architecture, type of xylem tissue, position and density of leaf stomata, photosynthesis and respiration rates, leaf nitrogen and phosphorus content, time of bud swelling, flowering period, and time of leaf fall. Functional traits number in the thousands. A global plant trait database exists at: https://www.try-db.org/TryWeb/Home.php. This database currently lists over 2500 traits with millions of data entries by vegetation scientists from around the world.

Nock et al. (2016) provide the following definition for functional traits: "Functional traits are morphological, biochemical, physiological, structural, phenological or behavioural characteristics of organisms that influence performance or fitness." For example, a trait may influence a species ability to colonize a habitat or to tolerate environmental changes. A trait may influence ecosystem properties, such as nutrient recycling. Nock et al. (2016) provide a general discussion of functional traits.

Plant traits used for ecological assessments may be distinguished according to their type of variability. Most traits vary "within species along environmental gradients, or in response to specific environmental changes" (Pérez-Harguindeguy et al., 2013). 'Stable traits' have low intraspecific variation compared to interspecific variation, while the reverse occurs with 'variable traits'. Examples of variable traits include vegetative height, onset of flowering, branching architecture, and spinescence. Stable traits include life forms, clonality, and dispersal modes (Pérez-Harguindeguy et al., 2013).

Plants in Space (Communities)

One can compare how two areas may be related by examining their floristic compositions. The flora of an area is represented by a list of species. If floras from two areas are similar, then it is likely that they have similar habitats. The vegetation in these areas can also be compared. Vegetation represents not only an area's flora but also its structure (e.g., absolute and relative species abundance, dominance, vegetative layering, etc.). This information improves and enlarges the functional descriptions of plant communities.

When confronted with the complexities of a landscape, it is helpful to formulate a basic approach that will allow comparisons of different sites. Two situations are usually encountered in the field. Floristic composition, plant abundance, distribution patterns, etc. may change with distance along a gradient. For example, a sharp zonation pattern of species may be evident leading down to a pond. Or a change in vegetation may occur as one moves up a mountain. If interest lies in examining the change in vegetation, then observations are centered over the transition area. A second situation occurs when interest is focused on comparing an apparently homogeneous region of the landscape to another homogeneous region. A homogeneous vegetated area contains a single physiognomy and has a consistent distribution of dominant taxa in each vegetation layer. It has no abrupt changes or obvious gradients in environmental factors (e.g., slope, aspect, soil, etc.). Dominance generally refers to those species with maximum influence on the habitat (see Mueller-Dombois and Ellenberg, 1974 for a discussion of "homogeneity" and "dominance"). Dominance can be estimated with varying criteria (e.g., the tallest species, the species with the greatest density of individuals, the species with the greatest canopy coverage).

Whether observing changes along a gradient or examining homogeneous areas, a basic spatial unit of observation is chosen. Select the location, shape, and size of spatial units (i.e., plots) to reflect what appears to be representative of the area of interest and appropriate for the plant sizes present. A transition area will require several plots, whereas a homogeneous region may be adequately described with one. Anomalies occurring outside of the plots can be separately noted.

A plot consists of an area of the ground surface within which various observations are made. The first step is to select an appropriate size for the plot. In quantitative work, plots are often laid out and marked on the ground. Different sized plots are evaluated in order to determine the appropriate size for sampling the vegetation. Then multiple, appropriately sized plots are laid out and sampled so that statistical tools can be applied to the data. Bonham (2013) provides detailed descriptions of plot techniques for the quantitative estimation of plant frequency, cover, density, and biomass. Time constraints on an inspection-level outing do not allow for such detailed work.

Since interest is centered on developing and recording a typical description of the vegetation, qualitative estimation methods are suggested. Observations are made over a predetermined ground surface, but the surface area is visually estimated rather than physically laid out. To select an appropriate size of the ground surface in which to make observations, refer to plot sizes used by others (see Table 2).

Convenient plot sizes are as follows. Sizes for sampling

trees (DBH of 5 inches and over) may be a circle with a radius of about 30 feet (roughly 10 m). DBH is the trunk diameter at breast height [approximately 4.5 ft (1.37 m)]. Climbing woody vines may also be analyzed in a 30-foot-radius plot. Saplings (trees more than 1 and less than 5 inches DBH) and shrubs (multi-stemmed woody plants usually less than 13 to 16 feet high) can be sampled using a smaller plot [e.g., a 15-foot (roughly 5 m) radius]. Subshrubs (low growing shrubs usually under 1.5 feet tall), graminoids, and forbs use yet smaller plots [e.g., a 5-foot (roughly 1.5 m) radius]. The woody seedling/sapling break is defined by the US Forest Service at 1 inch DBH/DRC (DRC is the diameter at the root crown). Seedlings may also be sampled in the smaller plot.

In order to evaluate community characteristics, several sample plots may be required. For example, the US Forest Service uses four subplots, each with a radius of 24 feet, to sample a site (Bechtold and Scott 2005). The total evaluation area is dependent on the type of vegetation and its variability over space. Braun-Blanquet (1932) suggested that "[i]n a Mediterranean therophyte community areas of 0.5 or 1 sq. m. are appropriate; in the beech or spruce woods at least 500 sq. m. must be taken for the tree layer." Commonly used plot sizes are shown in Table 2.

Keep in mind that the outing under discussion is for inspection purposes, and it is not suggested that plots actually be physically established. These sizes are provided simply as a frame of reference when establishing areas to evaluate. Select evaluation areas that are large enough to capture typical vegetative structure.

While there is no minimum area size for a plant community, it is obvious that a site with only two or three trees is not a forest. The trees would be incorporated into the variation of the surrounding community. Nevertheless, recognized plant communities can be very small, as, for example, in a small depression containing a vernal pool.

Table 2: Commonly used plot sizes (based on Coles-Richie, et al. 2015)

Approximate Plot Size m²	Circular Plot Radius m (ft)	Square Plot Side Dimension m (ft)
~50	4.0 (13.1)	7.1 (23.3)
~100	5.6 (18.5)	10 (32.8)
~400	11.3 (37.0)	20 (65.6)
~500	12.6 (41.4)	22.3 (73.3)
~800	16.0 (52.4)	28.3 (92.7)
~1,000	17.8 (58.5)	31.6 (103.7)
~2,500	28.2 (92.5)	50.0 (164.0)

4: Biotic Characteristics

This chapter deals with the types of observations used to evaluate plants and plant communities. There are, of course, hundreds of types of observations that could be made. The material below emphasizes plant and community attributes that are easily observed and useful for inspection purposes. (Some plant attribute determinations require access to the internet and may not be immediately available in the field.) An observer may have a particular interest that can not be evaluated by the techniques described below. In this case, the development or use of other methods is encouraged.

Several exercises listed under **Plant Observations** below suggest making observations of plant parts using plants found in particular strata. Strata are defined and discussed in the **Community Structural Attributes** section later in the chapter.

It is important to remember that all of our observations represent conditions at the time of an inspection. As habitats are evaluated and compared, clues may be seen that suggest both past and future conditions. Past conditions can be evaluated by reviewing photographs taken during previous years and decades. For example, sequences of Landsat photographs in visible color and in false color infrared have been available since 1972.

Plant Observations

This section provides discussions and observation techniques that apply to individual plants. An effort has been made to include a variety of techniques that will allow an appreciation of plants growing in many different habitats (i.e., sites of interest). The list of techniques has been constrained to those that are readily applicable during an inspection-level outing.

The Plant List

The first order of business on an inspection outing is to begin a plant list. Chapter 2 gave some sources whereby it may be possible to predict encountering a number of species, so check for those first. Key out unknown plants as time permits. If there isn't enough time, assign a number to the plant, then take photographs and, if allowed, appropriate specimens for later species determination.

Species observations are made by strolling through the area of interest. If community attributes will be evaluated during the inspection, avoid trampling plants within the plot examination area (see Chapter 4: **Community Structural Attributes**). Select adult specimens (plants capable of reproduction) for identification. Place emphasis on dominant and common species in each layer of vegetation. (**Avoid touching toxic plants such as poison ivy, poison oak, and poison sumac. Avoid touching stinging nettles and other skin-irritating plants**).

Identifying plants using a plant identification application on a smart phone can be helpful. To use in the field, a photograph of the plant is taken and loaded into the application. Access to the internet is required; however, a photograph can be used at a later time if access is unavailable. Photographs of the plant habit, closer views of bark, branches, leaves, and, if available, flowers, are recommended. Select clear, well focused photographs for input into the application.

Experience with a couple of these applications suggests that better results are obtained when one plant part is used compared to another, depending on the particular application. In a comparison of two applications, this author noted that identification accuracy ranged from 64% to 76%. Preexisting photographs (personal collection) of 50 different common species found in one region were used in the comparison.

Because of this, it is recommended that any result be evaluated by consulting a field guide. Read the plant's description and identify key characteristics to confirm the scientific name. Be cognizant of plant synonyms when comparing results from the application to the field guide. Although the accuracy of plant identification applications can be expected to improve over time, there will still be an advantage to using field guides since they give an overview of the species and place it in context with other species.

Exercise #1: Plant List
Stroll through the area of interest, identify and record plant species.

Growth Habit

The growth habit of a plant refers to its overall appearance, its characteristic form or morphology, its herbaceous or woody character, its size, etc. Pérez-Harguindeguy et al. (2013) state: "Growth form may be associated with ecophysiological adaptation in many ways, including maximising photosynthetic production, sheltering from severe climatic conditions, or optimising the height and positioning of the foliage to avoid or resist grazing by particular herbivores, with rosettes and prostrate growth forms being associated with high grazing pressure by mammals."

Table 3 lists habits identified by Pérez-Harguindeguy et al. (2013). Table 4 lists growth habits for mature (i.e., of reproductive age) plants found in the NRCS PLANTS Database.

Table 3: Growth Habits
(from Pérez-Harguindeguy et al., 2013)

• **Terrestrial, mechanically and nutritionally self-supporting plants** 　○ **Herbaceous plants**

- - Rosette plant
    - Elongated, leaf-bearing rhizomatous
    - Cushion plant
    - Extensive-stemmed herb
    - Tussock
  - Semi-woody plants
    - Palmoid
    - Bambusoid
    - Stem succulent
  - Woody plants
    - Prostrate subshrub
    - Dwarf shrub, or subshrub
    - Shrub
    - Tree
      - Excurrent
      - Deliquescent
    - Dwarf tree
- Plants structurally or nutritionally supported by other plants or by special physical features
  - Epiphyte
  - Lithophyte
  - Climber
    - Herbaceous vine
    - Woody vine, including liana
    - Scrambler
    - Strangler
  - Submersed or floating hydrophyte
  - Parasite or saprophyte

Table 4: USDA Growth Habits and Definitions
(From: USDA NRCS PLANTS Database, accessed 2023)

PLANTS Description	PLANTS Definition	Note
Forb/herb	Vascular plant without significant woody tissue above or at the	Applies to vascular plants only. Federal

PLANTS Description	PLANTS Definition	Note
	ground. Forbs and herbs may be annual, biennial, or perennial but always lack significant thickening by secondary woody growth and have perennating buds borne at or below the ground surface. In PLANTS, graminoids are excluded but ferns, horsetails, lycopods, and whisk-ferns are included.	Geographic Data Committee (FGDC) definition includes graminoids, forbs, and ferns.
Graminoid	Grass or grass-like plant, including grasses (Poaceae), sedges (Cyperaceae), rushes (Juncaceae), arrow-grasses (Juncaginaceae), and quillworts (*Isoetes*).	Applies to vascular plants only. An herb in the FGDC classification.
Lichenous	Organism generally recognized as a single "plant" that consists of a fungus and an alga or cyanobacterium living in symbiotic association. Often attached to solid objects such as rocks or living on dead wood rather than soil.	Applies to lichens only, which are not true plants.
Nonvascular	Nonvascular, terrestrial green plant, including mosses, hornworts, and liverworts. Always herbaceous, often attached to solid objects such as rocks or living or dead wood rather than soil.	Applies to non-vascular plants only, in PLANTS system this is groups HN (Hornworts), LV (Liverworts), and MS (Mosses).
Shrub	Perennial, multi-stemmed woody plant that is usually less than 4 to 5 meters (13 to 16 feet) in height. Shrubs typically have several stems arising from or near the ground, but may be taller than 5 meters or single-stemmed under certain environmental conditions.	Applies to vascular plants only.
Subshrub	Low-growing shrub usually under 0.5 m (1.5 feet) tall, never exceeding 1 meter (3 feet) tall at maturity.	Applies to vascular plants only.
Tree	Perennial, woody plant with a single stem (trunk), normally	Applies to vascular plants only.

PLANTS Description	PLANTS Definition	Note
	greater than 4 to 5 meters (13 to 16 feet) in height; under certain environmental conditions, some tree species may develop a multi-stemmed or short growth forms (less than 4 meters or 13 feet in height).	
Vine	Twining/climbing plant with relatively long stems, can be woody or herbaceous.	Applies to vascular plants only. FGDC classification considers woody vines to be shrubs and herbaceous vines to be herbs.

Exercise #2: Growth Habit

Growth habits are defined along a continuum, so be aware that intermediate forms may be encountered. Rewrite the plant list by grouping species into growth habits as defined in Table 4. (Refer to the habit definitions listed by Pérez-Harguindeguy et al. (2013) if using Table 3.)

Life Form

A life form is a growth form displaying a relationship with particular environmental factors (Mueller-Dombois and Ellenberg, 1974). The best known classification of life forms was developed by C. Raunkiaer in the early 20th century.

Raunkiaer was interested in the expression of vegetation with respect to climate on a global scale. Temperature and moisture were (and still are) considered controlling at this scale, and his aim was to divide the world into "equiconditional" regions.

Regions of similar vegetation are related by similar climates and composed of similar life forms (functional plant types). For example, succulent euphorbias from the deserts of southern Africa look very similar to our desert cacti. Through convergent evolution, both groups have developed succulent branching stems, longitudinal ribs, small leaves, short spines, shallow fibrous root systems, and crassulacean acid metabolism. These adaptations help plants function in hot, low-precipitation areas.

Raunkiaer thought that morphological, anatomical, and intracellular characteristics existed that allowed plants to survive the harshest, unfavorable conditions over the course of a year. Such unfavorable conditions were present when temperatures were too low or precipitation was insufficient. To categorize species, he chose to use plant life forms with structural features that enhance survival during the unfavorable season. He identified 30 life forms (Raunkiaer, 1907), but reduced this number to 10 when applying his life forms to the analysis of vegetation. Protection of meristematic tissue (perennating buds) necessary for regrowth after the cessation of adverse conditions was the primary consideration. Life form descriptions are as follows:

> Phanerophytes: primarily woody perennials (trees, shrubs, and lianas) with resting buds more than 50 cm above ground. (Raunkiaer used three categories of phanerophytes based on height: mesophanerophtes and megaphanerophytes over 8 m, microphanerophytes between 2 and 8 m, and nanophanerophytes under 2 meters.)
>
> Chamaephytes: woody and herbaceous plants with perennating buds close to the ground (a maximum of 20 or 30 cm above the surface).
>
> Hemicryptophytes: perennating buds at or near the soil surface (protected by the surrounding soil and withered remains of the plant).
>
> Cryptophytes: buds subterranean or underwater. (Raunkiaer used two categories of cryptophytes: geophytes with perennating buds in the soil, and helophytes/hydrophytes with buds at the bottom of water.)
>
> Therophytes: annual plants that survive cold or dry

seasons as seeds.

Raunkiaer hypothesized that life forms, when arranged from mega- and meso-phanerophytes, micro-phanerophytes, nano-phanerophytes, chamaephytes, hemicryptophytes, geophytes, helo- and hydrophytes, to therophytes, form a series with increasing adaptations for survival of unfavorable seasons (i.e., harsher temperature and/or moisture conditions).

Raunkiaer also added epiphytes and stem succulents to his list of life forms because they were "highly characteristic of certain floral regions" (Raunkiaer, 1908).

Exercise #3: Life Form

Add a column to the growth habit list. Assign species to one of the following:

> Phanerophyte,
> Chamaephyte,
> Hemicryptophyte,
> Geophyte,
> Hydrophyte,
> Therophyte,
> Stem Succulent, and
> Epiphyte.

Life Duration

Assuming adequate resources and an absence of externally influenced catastrophic events, plants complete their life histories (e.g., seed germination, vegetative growth, flowering, and death) within certain time periods, generally depending upon taxon. Non-clonal plants have limited lifespans. Clonal plants, such as many rhizomatous species, may have nearly unlimited lifespans.

Flowering and subsequent seed production are energy-intensive activities. In annuals, enough photosynthate needs to be produced within one year for seed production to occur. Biennials generally grow vegetatively the first year and only mature during the second. However, excellent growth can lead to a large enough biennial plant that some may flower during the first year. Perennials, such as trees, may require a number of years to reach sexual maturity.

Some species, particularly short-lived, faster-growing ones, may have different life histories depending on geographic location or habitat. As such, life history is best assessed in the field (Pérez-Harguindeguy et al., 2013). Plants with woody stems are easily identified as perennial. Herbaceous plants with perennating organs other than seeds live for two or more years. Biennial plants grow from a storage root the second year. To identify biennials in the field, look for individuals with a storage root but not an inflorescence. Other individuals will have both.

Plants that have specialized perennating organs but are not biennials would be perennial. Search for rhizomes, stolons, bulbs, etc. Perennial herbaceous plants lacking specialized perennating organs may resprout from their root-crowns. Look for wrinkles or scars on the crown from bud outgrowths in previous seasons. Over time, such plants can develop thick and even woody root crowns.

Annual plants have relatively soft and smooth roots. Root thickness extends continuously into the stem. (First year perennials will be similar.)

Exercise #4: Life Duration
Add a column to the growth habit list for life duration. Classify plant species as annual, biennial, or perennial. (Perennials can be categorized further. See Pérez-Harguindeguy et al., 2013.) Compare findings with durations found in the "Plants Database" (USDA, NRCS, 2023).

Phenology

Over the course of a year, plants grow and develop through a number of stages. Seeds from annuals germinate, seedlings develop, vegetative growth occurs, followed by flowering, and then seed dispersal. Resting perennials break dormancy and, depending on the species, may flower before vegetative growth or produce vegetative growth before flowering. Seeds are produced and dispersed either in the same year or the following year. Deciduous plants will eventually drop leaves at one time. Leaf senescence in evergreen plants will occur over a period of time, with new leaves replacing senescent ones, allowing the maintenance of green foliage throughout the year.

Phenology refers to the timing of recurring biological events (Koch et al., 2007). Environmental factors, such as temperature, moisture availability, and light, may affect phenology. Various local and global trends in phenology exist that suggest possible selective pressures are operating on the timing of plant life histories.

Phenology is constrained by a taxon's phylogeny, but within these constraints, both abiotic and biotic selective forces exist. A combination of selective pressures, such as seasonal climatic changes, resource availability, and the presence of pollinators, predators, and seed dispersers, may result in a compromise in the timing of life history phases. See Fenner (1998) for a review of phenology related to leafing, flowering, and fruit production by species and communities.

Phenological field studies require repeated visits, making them unsuitable for inspection outings. However, good practice requires recording phenological stages when visiting a site. The timing of these stages can be compared between sites if areas are visited on the same day, or at least within a few days of each other.

Various authors have developed plant phenological descriptors that differ in their details. For example, Braun-Blanquet (1932) used seven stages [i.e., in foliage, leafless, buds, flowering, fruiting, seedlings, and assimilating (photosynthetic)]. The British Columbia Ministry of Forests and Range and the B.C. Ministry of Environment (2010) identify vegetative stages for deciduous trees or shrubs (11 stages), conifers (7 stages), herbs (11 stages), ferns (11 stages), and grasses (11 stages). Generative stages for trees, shrubs, and herbs (14 stages), grasses (12 stages), and ferns (4 stages) are included.

The following principal stages (Koch et al., 2007) are used here.

- *Germination / sprouting / bud development*
- *Leaf development (main shoot)*
- *Formation of side shoots / tillering*
- *Stem elongation or rosette growth / shoot development (main shoot)*
- *Development of harvestable vegetative plant parts or vegetatively propagated organs / booting (main shoot)*
- *Inflorescence emergence (main shoot) / Heading*
- *Flowering (main shoot)*
- *Development of fruit*
- *Ripening or maturity of fruit and seed*
- *Senescence beginning of dormancy*

Exercise #5: Phenology

Add a phenology column to the growth habit list and record the stage(s) for representative specimens of each species.

Plant Height

Plant height is measured from the ground surface to the upper boundary of the main photosynthetic tissues but excludes inflorescences. The maximum stature that a typical mature individual attains has been associated "with growth form, position of the species in the vertical light gradient of the vegetation, competitive vigour, reproductive size, whole-plant fecundity, potential lifespan, and whether a species is able to establish and attain reproductive size between two disturbance events (such as e.g. fire, storm, ploughing, grazing)" (Pérez-Harguindeguy et al., 2013).

For plants with "major leaf rosettes and proportionally very little photosynthetic area higher up, plant height is based on the rosette leaves" (Pérez-Harguindeguy et al., 2013). "Epiphytes are measured from the upper foliage boundary to the center of the basal point of attachment" (Pérez-Harguindeguy et al., 2013).

A number of instruments and techniques exist for determining the height of trees and tall shrubs from the ground. While the concept of tree height appears simple, many difficulties and errors can arise when making field measurements. Techniques and potential measurement problems are discussed in a number of texts (see Leverett & Bertolette, DeYoung, 2016 and Natural Resources Conservation Service, 2004). For purposes of a site inspection, the basics of the traditional tangent method using a clinometer and pacing is presented below. The tree trunk from base to tip is assumed to be vertically aligned with the level distance line (see Figure 1). Thus, two right triangles are formed from which tree height can be determined when angles A and B, and distance to the tree are measured.

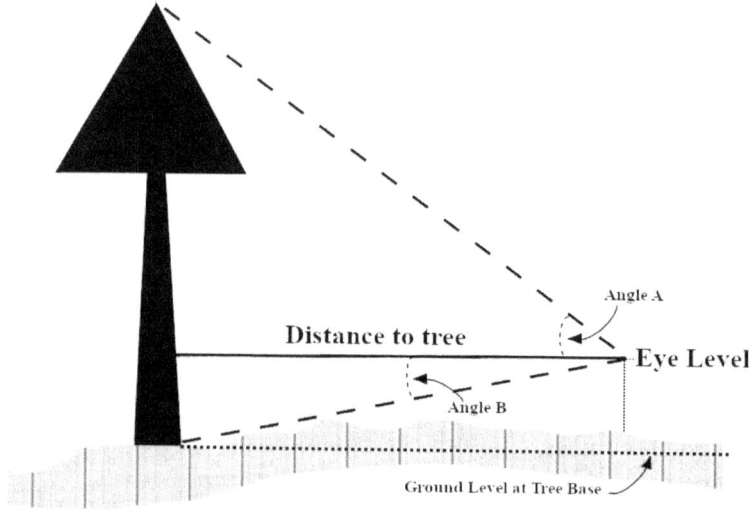

Figure 1: Tree Height Measurements.

Walk away from the tree to a point where both the base and top can be seen. Measure angel A and angel B. Pace to the tree to measure its distance. Note that if the terrain is steep, the distance measurement will not be along level ground and hence an error will be introduce into the calculated height. Therefore, try to find a sighting point along level terrain.

Determine the tangent of angle A and calculate the height above eye level using the following formula:

Height (above eye level) = tan(angle A) × Distance from tree

Determine the tangent of angle B and calculate the distance from eye level to the tree base using the formula below. (Note: eye level at the tree is lower than the eye level at the point of observation.)

Height (below eye level) = tan(angle B) × Distance from tree

Add the two results above to determine the tree height. If eye level at the point of observation is below the base of the tree, make a second measurement to the base as above, but then subtract for total plant height.

Assuming angles below the distance to tree line are negative, then a universal formula for tree height is:

Height = Distance × (tan A − tan B).

Even a **square** piece of paper can be used for rough height measurements. Fold the paper diagonally so that a right triangle is formed. The other two angles of the triangle are each 45°. Walk away from the tree so that when the triangle is held up to the eye with the bottom side in a horizontal position, the top of the tree can be sighted along the triangle's hypotenuse. Make sure the ground elevation is the same as the tree's ground elevation. The distance to the tree equals its height above eye level. It should also be noted that applications are available for use in smart phones that allow height to be determined.

Exercise #6: Plant Height

Use a tape measure or clinometer to determine the typical plant heights of mature (i.e., capable of reproduction) common species.

Tree Trunk Diameter

Diameter at breast height (DBH) provides an estimate of the size of trees within an area of interest. The diameter of a tree is measured at a convenient height (i.e., "breast height"). Breast height measurements vary depending on the country and the purpose of the measurement. Standard forestry practices in the US define breast height as 4.5 ft (1.37 m) above the uphill ground surface touched by the tree.

DBH can be used to determine dominance from basal area coverage, which is not considered for inspection outings because of the necessity of establishing plots or using optical instruments. [Dominance can also be estimated by canopy coverage (see Exercise #21 for estimating coverage)].

Leverett & Bertolette, DeYoung, (2016) and Natural Resources Conservation Service, (2004) provide various techniques and detailed instructions on measuring stem diameter. On an inspection outing, diameter can be determined with a tape measure. A 2 meter (6.5 ft) tape measure allows tree diameters of up to 0.6 m (1 ft) to be measured. Trees often exceed this limit; hence, plan on taking a longer tape measure if interested in this parameter. Alternatively, carry a length of string that will wrap around the largest trees expected on an outing. Use the string to encircle the tree at breast height, and then measure the length of the string in segments.

The concept of DBH appears simple, but many difficulties and errors can arise when making field measurements. If the tree is growing at an angle, measure the DBH at 4.5 ft along the trunk and perpendicular to the trunk axis. If roots are above ground, measure 4.5 feet above the root crown. But swelling, bulges, branch whorls, burls, or cankers may be present at breast height. In general, when a structure interferes with a measurement at breast height, measure the smallest circumference between 4.5 feet and the ground. A tree trunk may be split into two leaders, or two separate trees may be fused together. The trunk may be injured and only partly present at breast height. There are rules for these situations [see Leverett & Bertolette and DeYoung (2016)], but by choosing "representative" trees for measurements, these situations will likely be avoided.

After measuring DBH, calculate the tree diameter using the following formula:

Diameter = Circumference/3.14

Exercise #7: Tree Trunk Diameter

Measure the tree trunk circumferences at breast height and calculate and record the diameters of common, representative trees in the canopy and subcanopy layers of vegetation.

Vegetative Reproduction

As their basic unit of observation, ecologists often use the individual. This is particularly true in population ecology, where interest centers on recruitment, migration, mortality, age distribution, etc. Vegetative reproduction is common in plants, and it is often difficult to identify an individual. Underground roots and rhizomes can often produce new shoots, leading to distinct differences in life spans for root systems compared to shoots. Two closely spaced shoots may be the product of one root system, i.e., the shoots are genetically identical clones.

One approach is to distinguish between vegetatively and sexually derived plants, i.e., ramets and genets. A single physiological individual resulting from vegetative reproduction is referred to as a ramet. Ramets originating from a single seed are grouped as a single genet.

A way of dealing with this conundrum is to dig up the plants and determine if they arise from a common structure, such as a rhizome. All connected shoots can then be grouped as a single individual. This approach poses numerous disadvantages, not the least of which is its destructive nature. Another serious problem is that as growth and vegetative reproduction continue, connecting structures (e.g., rhizomes) may sever or die, leaving separated plants. Such plants live as physiologically separated individuals of the same genet.

Plant ecologists may simply make the assumption that a patch of closely spaced shoots represents a single genet, which may or may not be true. Another approach is simply to focus on shoots as the unit of observation. Yet another approach is to genetically sample shoots and determine individuals through laboratory analysis. (See Tsujimoto et al., 2020, for an example of this approach.)

The ability to clone may be an advantage by enabling the plant to expand horizontally, migrate short distances, increase its ability to exploit patches of resources, and increase its competitive vigor.

A plant's vegetative reproduction may fall into one or more of three categories. Pérez-Harguindeguy et al. (2013) identify these categories as:

> *(1) regenerative clonal growth, occurring after injury and normally not multiplying the number of individuals, as with resprouting from a lignotuber;*
>
> *(2) additive (also termed multiplicative) clonal growth, which can be either the plant's normal mode of multiplication or can be induced by environmental conditions such as high nutrient availability, and serves to promote the spread of the plant;*
>
> *(3) necessary clonal growth is indicated when*

> *clonality is required for the year-to-year survival of the plant, as with many plants that perennate from rhizomes, bulbs, tubers or tuberous roots and have no, or weak, seed reproduction.*

On an inspection outing, it is usually not practical to investigate clonality for tree species, since the required excavation is too large and time consuming. The same may be true for shrub species, although subshrubs may lend themselves to this type of destructive observation. Herbaceous plants and plants living in the ground stratum may lend themselves to the study of clonality.

Clonality in plants may be suspected if they are found in patches, although the absence of patches does not disprove the presence of clones. Sometimes saplings may be found grouped around a "mother plant," which again suggests cloning. An extensive, high-density network of plants may also suggest cloning.

Plant fragmentation can be an effective means of reproduction, as can layering. Some above-ground, clonal organs can be directly observed, including stolons, bulblets, and gemmae. Below-ground organs such as rhizomes, tubers, turions, bulbs, corms, tuberous roots, suckers, and lignotubers can promote vegetative reproduction.

Exercise #8: Vegetative Reproduction

Add a column to the growth habit list for vegetative reproduction. Search for perennating organs as identified above. For those species that reproduce vegetatively, identify which of the above three categories is applicable. The species' modes of vegetative reproduction can be found in the literature. Some useful sources are:

> Fire Effects Information System, [online]. U.S. Department of Agriculture, Forest Service, Rocky Mountain Research Station, Fire Sciences Laboratory (Producer). Accessed 2023: https://www.feis-crs.org/feis/.

R. M. Burns and B. H. Honkala. (tech. Coords.). Silvics of North America 1. Conifers; 2. Hardwoods. Publication: Agriculture Handbook 654, U.S. Dept. of Agriculture, Forest Service. Accessed 2023: https://www.srs.fs.usda.gov/pubs/misc/ag_654/table_of_contents.htm.

USDA Natural Resources Conservation Services. PLANTS Database. (particularly plant guides and fact sheets) Accessed 2023: https://plants.sc.egov.usda.gov/home.

Gill, J. D. and W. M. Healy (compiled and revised). 1974. Shrubs and Vines for Northeastern Wildlife. USDA Forest Service General Technical Report NE-9. 1974. Accessed 2023: https://www.srs.fs.usda.gov/pubs/3955.

Francis, J. K. (Editor). Wildland Shrubs of the United States and Its Territories: Thamnic Descriptions: Volume 1. Publication: General Technical Report IITF-GTR-26, U.S. Dept. of Agriculture, Forest Service. Accessed 2023: https://www.fs.usda.gov/treesearch/pubs/27005.

Spinescence

Some plants reduce or inhibit animal grazing by growing sharp spines, thorns, or prickles. Spines consist of modified leaves, leaf parts, and stipules and contain vascular tissue. Thorns also contain vascular tissue and originate as modified, sharp twigs or branches. Prickles do not have vascular tissue and develop from the epidermis or cork tissue. Pérez-Harguindeguy et al. (2013) state: "Different types, sizes, angles and densities of spines, thorns and prickles may act against different herbivores."

While some species always develop spinose structures, others are induced to form them as a response to herbivory and other types of damage, such as pruning or fire. These structures may densely cover organs, thus reducing heat or drought stress, as well as herbivory.

Exercise #9: Spinescence

Observe and record any species with spinescence. Record the relative density of spinescence (i.e., low, intermediate, or high).

Branching Architecture

The number of branches per unit of stem length can be used to assess plant architecture. Various advantages have been identified for both high-density and low-density branching (see Pérez-Harguindeguy et al., 2013). High density may reduce vertebrate herbivore damage by reducing animal feeding efficiency and denying access to plant organs. A high density of branching results in a large number of growing tips, ensuring continued growth after herbivores have removed some tips. A less branched habit can be an advantage in environments where fast growth in height improves plant survival and reproduction (e.g., in a fire-prone savanna, during the early stages of secondary succession). Species that live in low-light environments under forest canopies may be more branched per unit height than species living in full sunlight.

The apical dominance index (ADI) can be used to measure branching intensity. First, select a representative branch (a branch that reaches the outer part of the canopy). Designate this as branch "A." Work backwards from the tip of its longest-living terminal and identify branches arising from "A." Determine the first branch from "A" encountered that is leafless at its base but bears secondary branches with leaves. This position is the starting point for measurements. Measure the distance from the starting point to the branch "A" tip (line L in Figure 2). Count the number of living branches (ignore dead branches) following the starting point on branch "A" to the tip (1, 2, and 3 in Figure 2). The ADI is the ratio of branch number to measured length in meters. In the case of

Figure 2, L is equal to 0.09 m, and hence the ADI is 33.3. ADI can vary from zero (no branching) to >100 per meter (extremely ramified).

Pérez-Harguindeguy et al. (2013) state: "Like spinescence, branching architecture is a plastic trait that can differ within a species on the basis of browsing history, fire history, access to light, plant vigour or disease and even water stress. Branching architecture is also variable depending on the age and life history of the plant".

Exercise #10: Branching Intensity
Observe and record ADI for several common, representative plants from one or more vegetation strata.

Figure 2: Branching Intensity of *Ilex glabra*.

Leaf Size

Leaves, as the primary photosynthetic organs of most plants, provide a wealth of ecological information related to plant growth and survival. Over 400 leaf traits are listed in the TRY database. Pérez-Harguindeguy et al. (2013) describe techniques to measure sixteen such traits. Most leaf traits do not lend themselves to an inspection-level outing since they require laboratory analyses. Leaf area (LA) is considered here.

LA is the one-sided area of an individual leaf (usually expressed in mm^2). Leaf size is a highly variable interspecific trait. Globally, LA ranges in size from less than 1 mm^2 to over 10^6 mm^2 (Diaz et al., 2015).

Environments that are wet, hot, and sunny support a predominance of large-leaved species. Arid, hot, sunny environments typically support small-leaved plants. High latitudes and elevations are also correlated with small leaves (Wright et al., 2017). Heat stress, cold stress, drought stress, nutrient stress, and high-radiation stress select for relatively small leaves. LA may be influenced by allometric factors (e.g., plant size, twig size, etc.) within individual climatic zones. Ecological strategies related to nutrient stress and disturbance may also affect leaf size (Pérez-Harguindeguy et al., 2013).

Destructive techniques are often used to determine LA by harvesting the leaf, flattening it out between glass plates (with a scale included), and either scanning or photographing the leaf. Image recognition software can then be used to determine LA. This process does not lend itself to an inspection-level excursion. However, good estimates of LA can be made by multiplying leaf length (L) by width (W), and then by a correction factor (CF). Multiplying length by width gives the area of a rectangle enclosing the leaf, which is an overestimation of LA. As a rule-of-thumb, a CF of 2/3 is often used to estimate LA. More accurate estimates can be made by incorporating leaf shape into the CF (Schrader, et al., 2021).

Schrader et al. (2021) developed CF values for a variety of leaf shapes found in angiosperms and ferns. They examined over 3000 leaf images from 144 families and 780 species and subspecies, grouping them by leaf morphology following Ellis et al. (2009). Three schemes were presented, two of which are considered here.

The first scheme categorizes leaves as either lobed or unlobed. Schrader et al. (2021) state: "A lobe is a marginal projection whose apical sinus is incised by >25% of the distance from the projection apex to the midvein (Ellis et al., 2009). Lobed leaves include highly dissected fern fronds." An unlobed leaf simply has no lobes, as defined above.

CF for unlobed leaves = 0.69.
CF for lobed leaves = 0.53.

The second scheme breaks each of the above categories into more specific shapes as follows (after Schrader et al., 2021):

Unlobed leaves
  Elliptic: unlobed leaf with the widest part in the middle one-fifth and L:W<10:1.
  Obovate: unlobed leaf with the widest part in the distal two-fifths.
  Ovate: unlobed leaf with the widest part in the proximal two-fifths.
  Linear: unlobed leaf with L:W>10:1, regardless of position of the widest part of leaf.

Lobed leaves
  Palmately lobed: lobed leaf where the major veins of the lobes are primary veins that arise from the leaf base.
  Pinnately lobed: lobed leaf where the major veins of the lobes are formed by costal secondaries.
  Fern: highly dissected fern fronds.

CF for Elliptic leaves = 0.70.
CF for Obovate leaves = 0.66.
CF for Ovate leaves = 0.68.
CF for Linear leaves = 0.71.
CF for Palmately lobed leaves = 0.58.
CF for Pinnately lobed leaves = 0.53.
CF for Fern leaves = 0.41

The above definitions apply to simple leaves. Leaflets are

used when dealing with compound leaves. The fern category applies to an entire fern frond. CF values were based on blades (i.e., without stipules or petioles). Plants without leaves and three-dimensional leaves, such as those found in many gymnosperms (e.g., pine needles), are excluded from consideration here. Figure 3 illustrates examples of basic scheme 2 leaves.

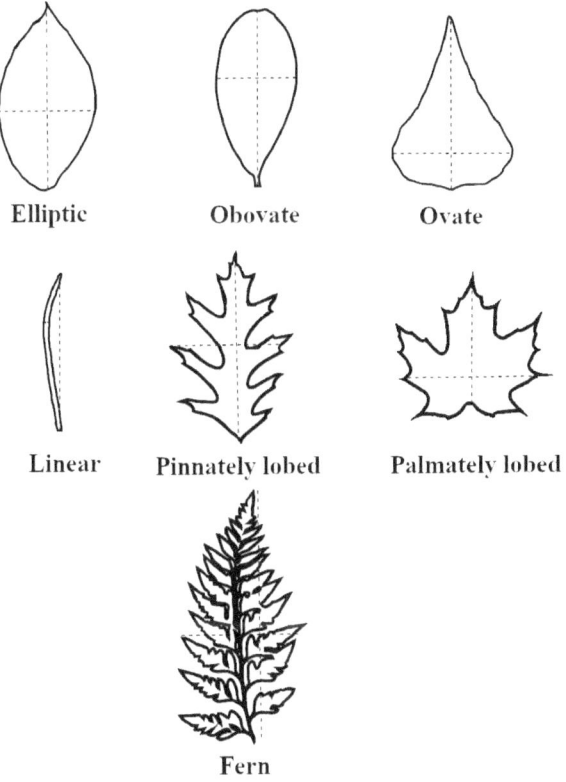

Figure 3: Examples of Leaf Shapes.
(based on: Schrader et al., 2021.)

Exercise #11: Leaf Size

Plants within the tree stratum are not included in this exercise because of the difficulty of accessing mid-shoot leaves. Select appropriate leaves from shrub and herb strata. For each species, observe and record five leaves from each of five plants. Use leaves that are mature but not senescent, and generally from mid-shoot heights.

Leaf selection criteria in the field depend heavily on projected data use. For example, comparing sun and shade leaves from a single site requires choosing exposed and shaded leaves (e.g., outer canopy leaves versus inner canopy leaves). Comparing leaf areas between physiognomically similar sites may require using leaves growing in similar light conditions (e.g., exposed mature leaves located on the outer mid-canopy). Comparing leaves between vegetation with dissimilar physiognamy (e.g., forest versus shrub communities) may require utilizing leaves from the same stratum (e.g., shrub layer).

Comparisons can be made between sites based on taxa or growth forms. The length, width, and leaf shape are noted for each leaf or leaflet measured. Leaf length is defined as the length of a **straight** line between the connection point of the leaf blade (e.g., point of connection to the petiole or, if sessile, to the stem) and the leaf apex. For grasses, record the length between the leaf apex and sheath. Leaf width is the longest extension of any two points on the blade edge **perpendicular to the leaf length axis** previously measured. (See dashed lines in Figure 3.) Choose a shape scheme identified above and record the leaf shape. In addition, record the number of leaves per node. Note the number of leaflets per compound leaf. Finally, calculate leaf areas and their average.

Leaf Retention

Woody plants may be classified as either deciduous or evergreen. Leaves of deciduous plants fall off during a short time span, such as in autumn or the start of a dry season. Evergreen plants have green foliage throughout the year. They also lose their leaves, just not all at once. Evergreen leaf-drop intensity may vary with the seasons. Some plants, referred to as semi-evergreens, exhibit an intermediate state. Foliage is persistent during a portion of an unfavorable condition, or it may be evergreen in mild climates but deciduous in harsh ones.

The USDA NRCS PLANTS Database lists species leaf retention for trees, shrubs, and subshrubs. Other growth habits are scored "No" by default for year-round retention.

Exercise #12: Leaf Retention

For trees, shrubs, and subshrubs, record the leaf retention characteristic. Make field observations during the "leaf-off" season or refer to leaf retention within the Plants Database under "characteristics".

Invertebrate Herbivore Damage

Herbivore and pathogen damage to plants represents perhaps the most significant challenge to vegetation throughout the biosphere. Plants have evolved a number of defense mechanisms to prevent or reduce damage from these agents. Freeman & Beattie (2008) offer a good overview of such defenses.

From the plant's perspective, herbivory represents the loss of living biomass, which is not necessarily a detriment to plant survival and reproduction. After all, deciduous plants lose their leaves each year, and evergreen plants also lose leaves, just not all at once. Many plants also self-prune. Lower tree branches die and fall off. (One advantage to this is a reduced risk of fire damage.) Herbaceous perennials generally die back each year, thus avoiding winter damage. From an energetics perspective, it costs the plant to produce organs; hence, one would expect the advantages of losing such organs to outweigh the disadvantages of lost energy investments.

When considering animal impacts on plants, the first thing that comes to mind may be wildlife grazing and browsing. Of course, animals that consume plants include a wide variety of mammals, birds, slugs, snails, and insects. Arthropods, particularly insects, are an important group relative to the consumption of living plant material.

The damage from leaf-chewing insects on plants becomes ever more apparent as the growing season progresses. A close inspection of the leaves of many species will reveal small portions missing, presumably from insect herbivores. In some instances, herbivory is so intense as to defoliate a tree. Most trees will survive a single defoliation, but repeated defoliation over a couple of years may result in plant death.

Exercise #13: Invertebrate Leaf Damage
Estimating the percent of a leaf damaged by chewing invertebrates offers a reasonably accurate and precise way of determining the severity of damage (Johnson et al., 2016). Select three typical plants from one or more vegetation strata and estimate the percent of leaf damage for three representative leaves per plant. Average percentages and record data.

Bark

Bark is generally described as a complex of tissues present external to the vascular cambium. Romero (2014) provides an excellent general review of the structure and ecological functions of bark.

Bark is essential for the continued life of woody plants and consists of the inner and outer bark. The inner bark contains active phloem elements that transport photosynthate throughout the stems and roots. Inner bark also contains parenchyma cells, fibers, and sclereids. Parenchyma cells provide a variety of functions, including radial transport (phloem rays), food storage, storage of tanins in vacuoles, and a source of meristimatic cells (ex., wound healing, origin of phellogen). Fibers and sclereids provide mechanical support. Inner bark may contain structures such as oil glands, laticifers, and resin canals.

The outer bark extends from the innermost tissue originating from the phellogen (cork cambium) to the stem surface. Commonly, new phellogens arise successively from the phloem parenchyma as the tree increases in girth. At maturity, all cells outside of the youngest phellogen are dead. These dead cells tend to have thick, ligno-suberized walls, making them resistant to water loss. Many species produce lenticels (porous tissue), which promote gas exchange between the air and inside stem tissue. The outer bark protects inner tissues from mechanical damage, heat (fire), cold, herbivore damage, and pathogens.

Tree survival and fitness may be influenced by bark structural characteristics. Trees with light-colored bark may be more resistant to radiation during leafless periods. Bark texture (smooth, scaled, fissured, laminated, cracked, or ridges) may also influence fitness (in Romero, 2014).

Bark contributes to the mechanical rigidity as well as flexibility of trees, thus providing protection against strong winds. It protects woody plants from herbivores through various barrier properties. For example, outer bark may be high in mineral compounds that may be abrasive to the mouthparts of herbivores. It may also contain exudates such as resins, gums, mucilages, or latex, which may discourage herbivores and protect inner tissues as wound healing progresses (Romero, 2014).

Bark provides insulation that helps to protect inner tissues during freezing periods or in the event of a fire. Thick bark may be associated with fire-prone areas and fire-adapted species.

Bark features affect the presence and distribution of arthropods (Romero, 2014). For example, deep fissures may provide refuge sites for insects. This, in turn, influences the feeding patterns of many insectivorous birds.

Perez-Harguindeguy et al. (2013) state: "Bark surface structure (texture) may determine the capture and/or storage of water, nutrients, and organic matter."

Twigs and young shoots may have photosynthetic capabilities (in the cortex or epidermis), and young bark may also contain tissue with chloroplasts. Bark characteristics tend to change as twigs grow. For example, the maturation of shoots is often associated with a loss of photosynthetic capability. However, a few species may retain this capacity, and stem photosynthesis can be a major source of photosynthate.

Over thirty functional bark traits are listed in the TRY Plant Trait Database. Bark traits discussed by Pérez-Harguindeguy et al. (2013) include bark thickness, the presence or absence of visible (liquid or viscose) gums or resins, bark surface structure, fissured stems, and bark investment (the ratio of bark thickness to stem radius).

Bark thickness observations requires injury to a woody plant. Injury ranges from cutting across the entire stem, to "blazing" from the stem surface into the wood, to using an increment borer. Small twigs and young branches may suffice when examining a shrub, and hence the injury may be minor.

When making observations on a tree, not only the young branches but also the thick trunk must be evaluated. Damage to the trunk can promote pathogen entry. For this reason, limiting observations to external trunk characteristics is suggested. These observations may include the use of a pocket knife to investigate bark characteristics near the trunk surface.

Observable bark characteristics include:

- presence or absence of visible (liquid or viscose) **gums or resins** in the bark (look at the bark surface and in shallow cuts);
- bark patterns (after Junikka, 1994)
  - **fissures** (i.e., longitudinal grooves between ridges)
    - **deep** (half or more of the thickness of the bark)

- **shallow** (less than half as deep as the bark thickness)
- **boat-shaped** (discontinuous oval or elliptical fissures)
- **short** (length is 15 cm or less)
- **long** (more than 15 cm long)
  - **ridges** (± continuous raised parts of bark between fissures)
    - **flattened** (the outer surface is plane in the cross section)
    - **hollow** (the outer surface is concave in the cross section)
    - **rounded** (the outer surface is convex in the cross section)
    - **V-shaped** (the outer surface is sharp in the cross section)
    - **reticulate** (the ridges join each other and irregularly divide, enclosing noncontinuous fissures);

- surface texture (after Pérez-Harguindeguy et al., 2013)
  - **smooth texture**
  - **very slight texture** (magnitudes of microrelief <0.5 mm)
  - **intermediate texture** (magnitudes 0.5–2 mm)
  - **strong texture** (magnitudes 2–5 mm)
  - **very coarse texture** (magnitudes >5 mm).

Exercise #14: Bark

Select representative woody plant species from one or more strata. Observe both the young and old bark for the characteristics noted above.

Dispersal Syndrome

From the plant's perspective, seed dispersal results in long-distance movement relative to that accomplished by many types of vegetative reproduction (e.g., from stolons or rhizomes). It is also more hazardous. Vegetative reproduction usually means growing close to a parent plant where adequate growth conditions have already been demonstrated. Given the nature of seed dispersal mechanisms, there is no control over where a seed will end up, and most seeds will not find their way to an appropriate site.

Nevertheless, plants exhibit a variety of dispersal mechanisms, which vary in the distance seeds may travel. Those species that produce heavy fruits and seeds may have their seeds deposited nearby, where conditions may be more appropriate for establishment than at sites further away. They may also have their seeds moved by birds and mammals to distant locations. Small, light seeds and seeds with specialized air buoyancy appendages tend to be dispersed considerable distances by wind, although the chances of arriving at a site with favorable conditions are low. Dispersal syndromes may be categorized as shown in Table 5.

Table 5: Dispersal syndromes
(after Pérez-Harguindeguy et al. 2013)

DISPERSAL TYPE	DESCRIPTION
Unassisted dispersal	Fruits and seeds fall passively without aid for longer-distance transport.
Wind dispersal (anemochory)	This category includes minute dust-like seeds, seeds with pappus or other long hairs, comas (trichomes at the end of the seed), flattened fruits or seeds with large "wings", spores of ferns and other cryptograms, and tumbleweeds (the whole plant or infructescence is rolled by the wind and releases seeds).

Internal animal transport (endo-zoochory)	Dispersules are eaten and dispersed by vertebrates after passing through the gut (e.g., birds, mammals, and bats). Dispersules include fleshy, often brightly colored berries, arillate seeds (seeds with specialized outgrowth that covers or is attached to the seed), drupes, and big fruits.
External animal transport (exo-zoochory)	Fruits or seeds become attached to animals (appendages on seeds, such as hooks, barbs, awns, burs, or sticky substances, may be present).
Dispersal by hoarding	Some mammals and birds may hoard brown or green seeds or nuts and bury them.
Ant dispersal (myrmecochory)	Dispersules have elaiosomes (specialized nutritious appendages) that make them attractive for capture, transport, and use by ants or related insects.
Dispersal by water (hydrochory)	Dispersules are adapted to prolonged floating on the water surface.
Dispersal by launching (ballistichory)	Some plants (e.g., *Impatiens*) develop restrained seeds that launch from the capsule when it opens.
Bristle contraction	Dispersules have hygroscopic bristles that promote movement with changing humidity.

Exercise #15: Dispersal Syndrome

Add a column to the growth habit list for dispersal and place species into the categories shown above. If seeds are not available for direct observation, consult the appropriate literature. Sources include local and regional floras and the following:

> Fire Effects Information System, [online]. U.S. Department of Agriculture, Forest Service, Rocky Mountain Research Station, Fire Sciences Laboratory (Producer). Accessed 2023: https://www.feis-crs.org/feis/.

Burns, R. M. and B. H. Honkala, (tech. coords.). Silvics of North America 1. Conifers; 2. Hardwoods. Publication: Agriculture Handbook 654, U.S. Dept. of Agriculture, Forest Service. Accessed 2023: https://www.srs.fs.usda.gov/pubs/misc/ag_654/table_of_contents.htm.

USDA Natural Resources Conservation Services. PLANTS Database. (particularly plant guides and fact sheets) Accessed 2023: https://plants.sc.egov.usda.gov/home.

Gill, J. D. and W. M. Healy (compiled & revised). 1974. Shrubs and Vines for Northeastern Wildlife. USDA Forest Service General Technical Report NE-9. 1974. Accessed 2023: https://www.srs.fs.usda.gov/pubs/3955.

Francis, J. K. (Editor). Wildland Shrubs of the United States and Its Territories: Thamnic Descriptions: Volume 1. Publication: General Technical Report IITF-GTR-26, U.S. Dept. of Agriculture, Forest Service. Accessed 2023: https://www.fs.usda.gov/treesearch/pubs/27005.

Species Wetland Status

Depending on a plant's tolerance to and requirement for water, it may be found growing in hydrophytic, mesophytic, or xerophytic environments. The terms are relative, so it is possible to speak of three habitat types being present in any particular landscape. For example, xerophytic habitats could be the tops of exposed hills or ridges; mesophytic habitats could refer to the middle slopes of hills; and hydrophytic habitats could refer to the often flooded depressions at the base of a hill.

Another method of characterizing the moisture affinity of a species is with reference to defined habitat characteristics.

This is achieved with respect to the wetter end of the habitat moisture spectrum (i.e., wetlands). The Federal Geographic Data Committee (2013) has defined a wetland as:

> *WETLANDS are lands transitional between terrestrial and aquatic systems where the water table is usually at or near the surface or the land is covered by shallow water. For purposes of this classification wetlands must have one or more of the following three attributes: (1) at least periodically, the land supports predominantly hydrophytes[1]; (2) the substrate is predominantly undrained hydric soil[2]; and (3) the substrate is nonsoil and is saturated with water or covered by shallow water at some time during the growing season of each year.*

Hydrophytes have been variously defined by different authors, although most have referred to plants growing in both the presence, at least during the growing season, of water and oxygen deficient substrates (see Tiner, 2012 for a discussion of definitions). The Federal Geographic Data Committee's (2013) definition is used here:

> *Any plant growing in water or on a substrate that is at least periodically deficient in oxygen as a result of excessive water content.*

A list of plants occurring in wetlands was published by the U.S. Fish and Wildlife Service in 1988. Responsibility for the National Wetland Plant List was transferred to the Army Corps of Engineers in 2006, and periodic updates and revisions have been made. Today's list may be found at: **NWPL Home v3.4-f9c (army.mil)**.

One of the unique features of the "National List" is its classification of plants within a region based on their fidelity to wetlands. Fidelity relates to the degree to which one finds a plant associated with a particular community and, hence, how good an indicator that plant is for the community.

The "National List" categorizes the fidelity of plants to wetland communities by assigning them to one of five major indicator categories based on how often they occur in wetlands versus uplands. Indicator categories for the "National List" are as follows (Lichvar et al., 2012):

> • *OBL (Obligate Wetland Plants). - Almost always occur in wetlands.*
>
> • *FACW (Facultative Wetland Plants). - Usually occur in wetlands, but may occur in non-wetlands.*
>
> • *FAC (Facultative Plants). - Occur in wetlands and non-wetlands.*
>
> • *FACU (Facultative Upland Plants). - Usually occur in non-wetlands, but may occur in wetlands.*
>
> • *UPL (Upland Plants) – Almost never occur in wetlands.*

If a species does not occur in wetlands in any region, it is not on the National List.

Exercise #16: Wetland Indicator Category

Add a column to the growth habit list and record the wetland indicator category for each species. A species' category may be found at the website listed above or on the "Plants Database" (USDA NRCS, 2023) site.

Wetland Plant Morphology

Some plants growing in saturated anoxic soils or under inundated conditions develop morphological, physiological, and/or phenological characteristics in response to such waterlogging. Tiner (1999, 2012) discusses a number of these characteristics. Table 6 identifies selected characteristics appropriate for an inspection-level outing.

Table 6: Morphological adaptations or responses to waterlogging
(Based on Tiner, 2012. Some characteristics may also be found in uplands.)

CHARACTERISTIC	DEFINITION WITH RESPECT TO WETLANDS
Stem hypertrophy.	Swelling of stem, particularly the lower portion. It is referred to as buttressing in trees.
Fluted trunks.	Base of trunk develops flutes.
Aerenchyma tissue in roots and other plant parts.	Soft tissue containing air spaces. Promotes internal air movement.
Hollow stems & roots.	Herbaceous plants, especially grasses, rushes, & sedges produce hollow stems. Promotes internal airflow.
Shallow root systems.	Root systems shallow to avoid anoxic conditions and locate oxygen. Very wide-spread wetland condition. (May form in uplands in very stony soils or where bedrock is close to surface.)
Adventitious roots.	Adventitious roots formed due to prolonged inundation. Roots usually formed near the water surface.
Pneumatophores.	Areial roots specialized for gas exchange (e.g., cypress knees).
Hypertrophied lenticels.	Enlarged lenticels found below to just above water line. Lenticels enhance gas exchange to/from internal tissues.
Heterophylly.	Submerged & emergent leaves with different leaf morphologies.

Exercise #17: Wetland Plant Morphology

Record any observed morphological adaptations or responses noted in wetland sites of interest.

Salt Resistance

Coastal environments and arid climate zones with poorly

drained soils typically contain areas with high concentrations of salt. These areas may support plants (halophytes) with specialized adaptations that reduce or avoid damage from excess salt. Pérez-Harguindeguy et al. (2013) identify three common strategies plants may use to tolerate these environments, i.e., selective root cation uptake, salt excretion, and salt compartmentalization. Of these three, only salt excretion lends itself to identification during field inspections.

Exercise #18: Salt Resistance

Salt excretion occurs through special glands or bladders that are usually located on the lower surfaces of photosynthetic organs. Select as many species as time permits. Using a hand lens, search for white spots (excreted salt crystals) on these glands. To confirm the presence of salt, lick the surface for a salty taste. Also check for salt excretions on the surfaces of the roots. This search is best done after a dry period, since salt will wash off with rain.

Community Structural Attributes

Plant communities (i.e., types of vegetation) are groups of species that occur in three-dimensional space in our landscapes. They can be classified into types that repeat, at least in broad detail, across the landscape. When environmental conditions change over time or space, plant communities change. Plants and plant communities function as foundation units, providing the energy for higher trophic levels; they are an integral part of ecosystem cycling (e.g., carbon, nutrient, and water cycling) and have a major influence on habitat characteristics.

Plant communities can be described based on their flora and/or their physiognomy. Physiognomy refers to the general appearance of vegetation (e.g., forest, savanna, grassland, etc.). Often, one to three dominant species and some notable environmental feature(s) are included in their names (e.g., Yellow Oak Dry Calcareous Forest).

Plant communities exhibit specific characteristics that can be noted and compared between sites. Strata, plant abundance, dominance, and dispersion (sociability) characteristics are discussed in this section. The evaluation techniques discussed below are suitable for inspection-level outings, although alternative quantitative methods exist for research requirements.

Some evaluations are based on ocular estimations relative to plot sizes. Plots are not physically established but rather visualized on the ground. Choose a site that appears typical for the community. Each plant strata is evaluated separately on the appropriate-sized plot (see Chapter 3: "Plants in Space" for a discussion of plot sizes). Plots may be nested inside one another and of varying shapes. Figure 4 illustrates various plot shapes and nesting patterns. For a reconnaissance-level outing, choose a shape that allows the plot to remain inside a homogeneous community. Plot sizes should be appropriate for the particular community and stratum under investigation (e.g., large enough to capture most of the species).

Identify a center point for a circular plot or a corner for a square or rectangular plot. Pace off a radius or plot side and visualize the area covered by the plot. For a circular plot, the pacing distance can be visualized in all directions from the center point to capture the area of investigation. For a square or rectangular plot, visualize the location of each plot side. For nested plots, evaluate the smallest plot size (i.e., the ground layer) first. This prevents trampling low-growing plants prior to evaluation.

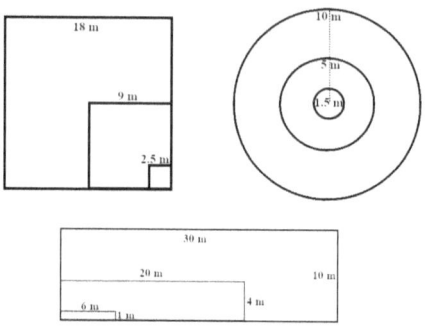

Figure 4: Nested plot designs

Strata Characteristics

When examining an area of interest, the vegetation can be visually divided into strata. Strata refers to the height ranges of foliage (majority of photosynthetic organs) that can be observed clustered at various heights above the ground surface. The number of strata may vary by plant community.

Layering, Abundance, and Coverage

For inspection purposes, complex communities (e.g., forests) may consist of the following: a tree layer (5" dbh and larger woody plants above 13 to 16 ft in height); a shrub layer (woody plants below 13 to 16 ft and above 1.5 ft in height); a ground cover layer consisting of subshrubs (woody plants less than 1.5 ft tall) and herbaceous plants (all forbs, herbs, and graminoids above 2 inches in height); and a ground stratum (small seedlings, mosses, lichens, etc.). Observations of the ground stratum often include non-living components of the ecosystem, such as dead and decaying leaves, twigs, and fallen trees. Vines and epiphytes are put in their own layers. The tree layer often consists of more than one stratum [e.g., main tree canopy, subcanopy, sappling layer (small trees less than 5" but more than 1" dbh)]. Sometimes a few trees, called emergents, can be seen that extend well above the general canopy layer. These may represent trees that are either much older or have especially fast growth rates compared to those in the general canopy layer.

Exercise #19: Strata

Identify the strata present in the area of interest and record their range of heights. Identify the species present within each strata. Since plants generally begin life on the ground, lower strata may contain species found in higher strata. Therefore, a species may be listed more than once because it is a constituent of more than one stratum.

Exercise #20: Abundance

Density refers to the number of individuals of a species per unit area or volume. Making an accurate determination of density is beyond the scope of an inspection outing; however, a rough estimate of density, called abundance, can be used. Visual inspection allows a species to be placed into one of five classes: (1) very rare, (2) rare, (3) occasional, (4) abundant, or (5) very abundant. Estimate and record the abundance of each species in each stratum. Note especially seedlings and saplings, since their presence suggests potential future conditions of the vegetation.

Exercise #21: Species Coverage

One vegetational characteristic of interest is the amount of plant material a species contributes to the community, i.e., a measure of the species' influence on other components of the ecosystem. Coverage (among other measures) is commonly used to arrive at an estimate of this factor and can be used to estimate species dominance. More than one type of coverage exists; however, the following definition is used here: "coverage is taken as an expression of the percentage of the ground included in a vertical projection of imaginary polygons drawn about the total natural spread of foliage of the individuals of a species" (Daubenmire, 1968). This statistic provides an expression of the two dimensional size of plant canopies but omits the vertical dimension. Analyzing coverage in each vegetational stratum (e.g., tree layer, shrub layer, etc.; see above for details of strata) lessens the vertical dimension problem.

To estimate coverage on an inspection outing, choose a site that appears typical for the community. While the details of coverage evaluations vary among authors, the following rapid method works well: Each plant stratum is evaluated separately on the appropriate sized plot, beginning with the lowest stratum.

The following discusses the tree layer as an example. Identify a center point and imagine a circular plot with a radius of about 30 feet. Next, estimate the percent of the imaginary plot covered by each species of tree. All foliage impinging on the imaginary plot is evaluated for each species. Overlaps of foliage or gaps in the canopy of individuals of a single species are ignored, i.e., concern is over the outer boundaries of each species' crowns. Assign each species in the layer to one of the classes in Table 7.

Table 7: Values for a coverage estimation technique (after Daubenmire, 1968)

Coverage Class	Range of Coverage, %	Midpoint of Coverage Class, %
1	0-5	2.5
2	5-25	15
3	25-50	37.5
4	50-75	62.5
5	75-95	85
6	95-100	97.5

After Daubenmire, 1968.

Note that the range of coverage is small for the first and sixth classes. Daubenmire (1968) explained this as follows:

> *Class 1 is made small so as not to overestimate poorly represented species, for there tend*

> to be many plants with small representation, and few with large. Class 6 is necessarily small since the midpoint of even this small range would give a plant which covered all or every plot, an average of only 97.5 percent.

On this topic, Mueller-Dombois and Ellenberg (1974) state:

> One may wonder why most scales show unequal class intervals. The main reason for this is a practical one. The chosen scale values allow for an easier estimation of species-cover-to-area relationship than do equal intervals of cover. Also, the less abundant species or species with small cover may sometimes have an important diagnostic significance, which requires a finer breakdown in the lower scale values as compared to the larger scale values.

Coverage classes lead to a convenient way to estimate coverage for each species. Judge whether or not the total species crown foliage area contributes more than 50% coverage over the plot. If the answer is no, estimate whether or not the species coverage is above 25%. If the answer is yes, then the species belongs in class 3. If not, then judge whether coverage is above 5%. If the answer is yes, then class 2 is chosen. If not, then the species belongs in class 1. The same process is used for species coverage that exceeds 50%. Judge whether or not crown foliage area exceeds 75%. If the answer is no, then choose class 4. If it exceeds 75%, judge whether or not coverage exceeds 95%. If the answer is no, then choose class 5. If yes, then choose class 6.

It is also helpful if the plot visualized is divided into halves and quarters. Then an estimate is made as to whether or not the total species crown cover will fit into one of the subplot areas.

Record coverage classes for each species in each layer. To

combine values in a stratum, use midpoints. Since different species of foliage may overlap, it is possible to have total estimates in a layer exceed 100 percent.

Exercise #22: Layer Coverage

Coverage concepts can also be extended to each stratum. Here, interest is centered on how much of the ground is covered by all vegetation. The same technique for species coverage is used, but all vegetation in a layer is considered instead of each species' coverage. In this instance, coverage can never exceed 100 percent since all foliage overlap is ignored.

Canopy coverage can be used to distinguish various physiognomic communities. For example, Swain (2020) uses coverage categories to distinguish between forests, woodlands, and shrublands, as shown in Table 8.

Table 8: Physiognomic categories

Community	Woody Coverage %
Forest	> 60
Woodland	25-60
Shrubland	> 25 shrubs; <25 trees

Physiognomic categories based on cover are arbitrary and may differ by author. For example, scrub/shrub wetlands are wooded wetlands where scrub/shrubs are dominant and have at least 30% aerial coverage of scrub/shrubs. Trees are dominant in forested wetlands and have at least 30% coverage of trees (Federal Geographic Data Committee, 2013.)

Estimate the canopy closure for each stratum.

Gaps and Shading

Through phenotypic plasticity and genotypic variation, each species is able to tolerate a range of light conditions. Despite this tolerance, species vary in their capacity to grow and survive under different amounts of shading. The minimum light required for a species's survival, i.e., its shade tolerance, is affected by a number of factors, such as ontogeny or environmental complexity. Shade tolerance can affect responses to environmental stressors such as suboptimal temperature and moisture conditions. (See Valladares et al., 2016, for a review of ecological factors associated with shade in the understory.) Shading is an important ecological factor that is itself affected by vegetation.

Plant structure and height influence shading in complex ways. Tall plants will obviously shade shorter neighboring plants. Individual shoot architecture as well as vegetation structure are important determinants of the understory and ground cover light environments. Space between overstory plant canopies allows light to reach further down toward the ground in these openings. Plant canopies themselves allow various amounts of light through to the ground (canopy openness) as diffuse light and sun flecks. Because sun flecks move as the earth rotates through the day, they can thoroughly illuminate the forest floor and play an important role in ground cover growth.

Exercise #23: Canopy Openness

Canopy openness can be measured by various means, such as a spherical densiometer, canopy photography, or by direct measurements of solar radiation (see Russavage et al., 2020 for an evaluation of these three methods). For inspection purposes, simply note whether the light environments under the tree or lower strata are well-lit (i.e., without over-topping vegetation), shaded, or highly shaded (dense shade).

Exercise #24: Canopy Gaps

Layer coverage (see Exercise #21 above) can provide information on sub-strata exposure to direct sunlight. A canopy gap, in the traditional sense, refers to an opening between plant canopies that arises through disturbance (senescent tree falls, wind, severe insect damage, fire, etc.). Muscolo et al. (2014) provide a review of the role canopy gaps play in ecosystem dynamics.

Identify and inspect any canopy gaps in the area of interest, and estimate their dimensions. Identify plant regeneration near the edge of a gap and also at its center.

Plant Distributions Within a Community

Plants may be distributed within a community in a variety of ways. For example, in walking through the spatial expanse of a community, a species may be irregularly observed or found almost everywhere. Species frequency can be used to indicate this type of distribution. Daubenmire, (1968) states: "It is defined as the percentage of occurrence of a species in a series of samples of uniform size contained in a single stand, the numbers and sizes of plants in each sample being ignored." One means of obtaining frequency data is to simply record the presence of a species within a number of plots. For example, if one hundred plots are established within a community and a species is present in 70, then it has a 70% frequency. Since plot size will affect frequency, an appropriate plot size must be determined. Frequency requires a relatively large number of plots. While an interesting statistic to have, frequency will not be further discussed because of the limited time available on an inspection outing.

Dispersion (sociability) represents another type of spatial plant distribution. Daubenmire (1968) states: "Whereas frequency expresses the relation of individuals to different parts of space, *sociability* (or *dispersion*) expresses the relation of individuals to each other." On a dispersion continuum, plants may be uniformly spaced in relation to one another (e.g., in an orchard), randomly spaced, or clumped. The type of dispersion may suggest a variety of interactions. For example, clumped dispersion may indicate vegetative reproduction. It may also suggest favorable microhabitats. A uniform distribution could indicate competitive factors are at play. Daubenmire (1968) discusses a number of factors that may produce various patterns.

Exercise #25: Dispersion

Various quantitative methods exist to determine dispersion; however, in keeping with time constraints of an inspection outing, the following qualitative method is presented. Estimate the dispersion class for each species based on Table 9.

Table 9: Dispersion Classes
(after Braun-Blanquet, 1932)

Class	Description
1	growing singly
2	slightly grouped
3	in small patches
4	in large patches
5	in an essentially continuous population

5: Abiotic Characteristics

Non-living characteristics of the habitat may impose conditions that inhibit or encourage the presence, abundance, and distribution of species in a community. Many of these attributes are easily seen or measured during inspections. Some require appropriate interpretations of maps. Others require obtaining data from internet sources. They provide ways to evaluate growing conditions, and provide information on insolation, shelter, exposure, soils, water availability, toxicity, and nutrient availability.

Landscape and Landform Features

Landscape, as used here, refers to a broad geographical area (e.g., dune fields, mountain ranges, outwash plains, and bolsons) made up of one or more types of related landforms. A landform is a local attribute that is more specific in its reference to conditions on the ground. For example, a site's location may be described as being on the leeward side of a mountain range in a depression. This site may receive comparatively less precipitation than a windward site but more moisture because of its low topographic position. Therefore, location within the landscape (e.g., mountain range) and landform (e.g., depression) are important ecological parameters. It should be noted that "landform" is used by some authors to refer to very large regions. The distinction between landscape and landform made here is to comply with definitions found in "Title 430 – National Soil Survey Handbook Part 629 – Glossary of Landform and Geologic Terms, Subpart A-Exhibits" (U.S. Department of Agriculture). A perusal of Subpart B of this document (629.10) will demonstrate the large number of recognized landforms.

Location

To determine location, one must know a site's latitude, longitude, and elevation. Location affects the site's weather conditions, such as precipitation, maximum and minimum temperatures, growing days, and length of the growing season. Higher elevations and hill summits are often exposed to colder and windier conditions compared to lower elevations. Many other characteristics, such as water availability and soil characteristics, also change with elevation.

Specific information on a location is available on maps and GPS. General information on growing conditions at a specific location is available from resources identified in Table 1 in Chapter 2. More local information may be derived from the observations discussed below.

Exercise #26: Site Location
Record the latitude, longitude, and elevation of sites of interest. Provide a brief description of landforms (see **Landforms and Microfeatures** below) and site positions within landscapes and landforms. Describe the best access routes and any features that will help relocate the site.

Insolation

Light intensity and duration at ground level are functions of time, season, latitude, weather, terrain, and canopy architecture. Canopy characteristics are discussed in Chapter 4: **Community Structural Attributes**.

Exercise #27: Shading
Note time, season, and cloudiness when making a site observation. Identify shading by mountains, hills, boulders, etc. When doing so, note the position of the obstructing object relative to the site so that its influence during the course of a

day can be estimated.

Exposure

Exposure, as used here, refers to how readily plants may be subjected to various environmental elements, generally with regard to negative impacts from meteorologically induced agents. Plants may be exposed to high winds, high or low insolation, wave action, salt spray, increasing soil salinity, cold air drainage, frost, fires, and other conditions. Negative effects on plants are generally related to the intensity, periodicity, and duration of exposure. Wind, insolation, and salt (including non-meteorologically induced soil salinity) are considered here; however, other types of exposures are discussed under **Landforms & Microfeatures** below.

Wind

Wind is a pervasive element in terrestrial ecosystems, and high winds may topple plants, break branches, result in shoot desiccation, and other effects. Plants have evolved to both withstand and take advantage of wind. For example, root and shoot habits affect susceptibility to wind throw; wind may mediate both pollen and propagule dispersal. Forest openings caused by windthrows offer microhabitats suitable for seed germination, establishment, and sapling release.

By reducing the boundary layer along a leaf surface, air movement may increase water loss through stomata. If the plant's water balance suffers and stomata close, photosynthesis may decline. Exposed areas are vulnerable to higher winds, which may increase water loss, while protected areas decrease it.

A reduced boundary layer increases the temperature gradient between leaf surfaces and the atmosphere. Hence, wind increases heat flow to and away from plants.

Wind also affects plants through other environmental factors. For example, an onshore wind brings humidity and the moderating temperatures of oceans and lakes inland. Wave action is intensified in exposed locations, particularly when the overwater fetch is substantial. Wind moves sand and other soil particles, which may cover and suffocate plants. It can move salt spray inland and can result in snow accumulation around plants, which may protect them from damaging cold temperatures.

Wind exposure decreases near the ground, so taller plants are more susceptible to wind. Trees block wind, so interior forest habitats are less subject to wind effects. Open, treeless terrain allows the wind to reach maximum influence.

Exercise #28: Wind Exposure
For an inspection-level observation, it may be sufficient to note the presence or absence of protected or exposed locations, the presence of wind screens (e.g., tall, thick growth of trees, hills, boulders, etc.), and the general openness of the site. Also note the direction from which exposure occurs and the fetch of water or land contributing to the exposure.

Salt

Exposure to high salt concentrations in soil and water can adversely affect plant growth, reproduction, and survival. Only species adapted to these environments may be found. Adverse reactions to salt may stem from a variety of reasons, including osmotic relationships that inhibit water absorption, salt accumulation on shoot surfaces, and high toxic metal concentrations. Saltmarshes along coastlines owe their existence to high salt exposure from tides. Salt concentrations decline as seawater becomes diluted by freshwater sources. Freshwater seeps along coastal shore environments may cause an abrupt change in species composition, and different communities may be recognized along the shoreline as one advances inland along an estuary.

High salt concentrations can play a major role in plant growth and species composition when encountered inland.

Lakes with only an inlet but no outlet are concentrating systems. Salt is brought into the lake via the inlet and concentrated by evaporation. Eagle Lake in Northern California is an example of such a system (although an artificial drainage system was constructed in the early 20th century, exposing low lying previously inundated land). Saltgrass (*Distichlis spicata*), a common species in coastal saltmarshes, can be found growing with high desert plants at Eagle Lake within the low lying but dry areas. Desert playas (pans, flats, dry lakes, flat-bottom depressions) found in deserts also accumulate salt and support unique vegetation.

Exercise #29: Salt Exposure
For inspection-level information, note the presence of any nearby salt sources, including salt from roadside spray and drainage.

Landforms and Microfeatures

A landform refers primarily to the shape of the terrain, such as hills, basins, flats, and other features. Examples of specific types of landforms include dunes, playas, kames, eskers, kettles, talus slopes, ridges, slopes, valleys, floodplains, and many others. Each landform has topographic features that result in specific and recurring combinations of characteristics, such as soils, microclimate, light, water transport, and water storage. Topographic features range in size from large hills and long slopes to microrelief consisting of small areas with relatively small elevational differences (relief in the order of 3 to 10 meters). Microfeatures generally refer to even smaller elements such as vernal pools and pit and mound topography. These topographic attributes influence plant growth, species composition, abundance, and distribution.

Slopes affect sun exposure, water and air drainage, temperature, and soil properties. The erosive power and sediment carrying capacity of both water overland flow and channel flow increase with increasing slope. Water infiltration declines, as does the length of time that soil water is present.

Gravity impinging on slopes affects other ecosystem characteristics. Cold air masses tend to flow downhill, collecting in low areas. Soil depth and organic content tend to decrease with increasing gradients, thus affecting nutrient budgets. A variety of slope characteristics are worth noting, including aspect, gradient, complexity, and length.

A consideration of landforms and microfeatures helps in interpreting combinations of environmental factors found in the landscape. As an example, given a closed basin at the base of a hill, one may suspect plants within the basin to be subjected to ponding or high ground water and damaging spring frosts. This low position may have deeper and finer-textured soil with relatively more organic matter and be exposed to less wind and higher humidity compared to a slope or summit location.

Extensive information on landform characteristics may be found in the "Soil Survey Manual, Handbook 18" (Soil Science Division Staff. 2017). Much of the following material has been obtained and summarized from this source.

Aspect

Aspect refers to the direction a slope faces. It is determined by taking a compass bearing directly down the slope. A bearing is recorded in degrees after accounting for declination. It may be sufficient to record a direction as a compass point (e.g., north, northeast, south, southwest, etc.). The direction a slope faces influences factors such as the winds received, evapotranspiration, temperature, and snow accumulation. The USDA Handbook 18, Soil Survey Manual, states:

> *Aspect can substantially impact local ecosystems. The impact generally increases as slope gradient and latitude increase. In the mid lati-*

tudes of the conterminous United States, this effect becomes particularly important on slopes of approximately 6 to 8 percent or greater. Increased or decreased solar radiation on slopes due to aspect can affect water dynamics across a site. In the northern hemisphere, north-northeast aspects reduce evapotranspiration and result in greater soil moisture levels, improved plant growth and biomass production, higher carbon levels, and improved drought survival rates for plants. Increased solar radiation on south-southwest aspects increases evapotranspiration and decreases biomass production, seedling survival rates, and drought survival rates for plants (Soil Science Division Staff. 2017).

Exercise #30: Aspect
Determine and record the aspect of the area of interest with a compass, smart phone, or map.

Flats

Areas with zero or very low grades can be termed flat and range in size from large coastal and continental plains to very small areas. The following emphasizes local relief.

Relatively large flat areas exist along the coasts in the form of saltmarshes. Basins, floodplains, dry lake beds, and large wetlands may be relatively flat. Vernal ponds have flat bottoms that may only be sparsely vegetated or be covered in zones of vegetation. Lakes and ponds may have seasonally inundated shelves along parts of their shoreline, which can support a variety of plant species. Seasonal ponds may provide herbaceous plant habitat. Flat, wooded swamps or herbaceous wetlands can occupy large areas.

Subtle and small changes in topography and elevation of only a few centimeters can result in very large differences in species abundance and distribution in flat areas.

Exercise #31: Flats
Look for and record plant zonation and inundation patterns in flat areas.

Inclines

The inclination of a slope is based on a line oriented up and down the slope. Slope gradient is the change in elevation between two points along the line divided by the horizontal distance between the points, and is expressed as a percent when multiplied by 100. It is convenient to rate slope gradients in classes as in Table 10. Slope may be determined with a clinometer, although some clinometers may give results in degrees above the horizontal. Table 11 provides some equivalencies between slope and angle.

Table 10: Definitions of Slope Classes
(From USDA Soil Survey Manual, 2017)

Classes for—		Recommended slope (gradient) class limits	
Simple slopes	Complex slopes	Lower (percent)	Upper (percent)
Nearly level	Nearly level	0	3
Gently sloping	Undulating	1	8
Strongly sloping	Rolling	4	16
Moderately steep	Hilly	10	30
Steep	Steep	20	60
Very steep	Very steep	> 45	

Table 11: Some Slope and Angle Equivalencies

SLOPE (%)	ANGLE (°)
0	0
1	0.57
3	1.72
4	2.29
8	4.57
10	5.71
16	9.09
20	11.32

30	16.7
45	24.23
60	30.96

Slopes can be divided into parts, as illustrated in Figure 5. The summit and shoulder of the simple slope are typically the driest parts, while the foot and toe of the slope are more moist. Soils on the backslope tend to be thinner compared to soils at the base of slopes. While backslope areas drain quickly, springs and wet areas may be present depending on subsoil conditions. Toeslopes can range from upland to wetland conditions; erosional processes result in more organic matter accumulating in depressions.

Figure 5 also illustrates what is meant by complex slopes. Topographic features on a complex slope may vary in size and complexity, leading to a wide range of microhabitats. Complex slopes are characterized by a number of steps and benches. Slope length can be maximized by simple slopes, while complex slopes reduce slope length. The localized breaks in a complex slope alter slope wash processes and often correspond to changes in soil types. Slope complexity influences the amount and rate of runoff, sediments carried by runoff, and soil temperature resulting from local variation in aspect.

Slopes can be categorized as convex, linear, or concave. The simple slope shoulder in Figure 5 is convex, the backslope is linear, and the footslope is concave. Slope shape refers to the curvature of the ground surface. This requires the feature to be evaluated in two paired directions, up and down slope and across slope, and results in nine possible combinations of the convex, linear, and concave shapes (see Figure 6).

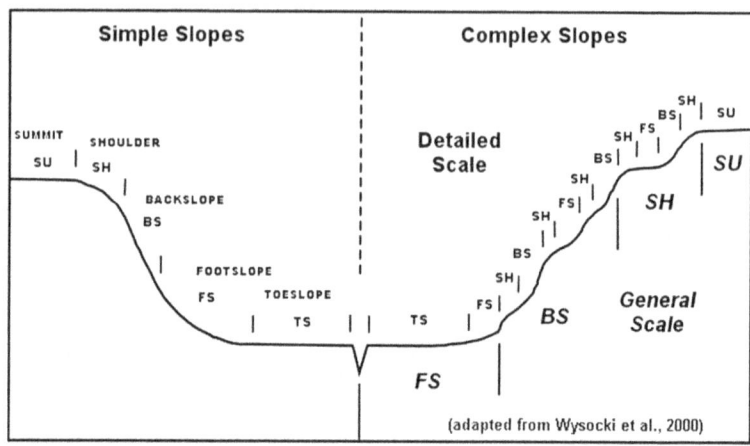

Figure 5: Slope Complexity.
(From USDA Soil Survey Manual, 2017.)

Features such as ridges, hills and knoll tops are convex. These shapes tend to drain off water, producing a drier environment. High convex features may be relatively exposed and windy. As radiation cooling occurs at night, the air mass becomes colder and heavier. It tends to drain off, leaving higher temperatures in place. The cold air mass moves down the gradient and can accumulate in concave depressions and basins, producing a local inversion (increasing temperature with elevation). These can become frost pockets in late spring and early fall, producing significantly colder temperatures and shorter growing seasons compared to areas higher up.

Microfeatures occur on most landforms and consist of very small variations of a few feet or even a few inches in topography. Despite their small size, they can have substantial impacts on plant distribution and abundance. In flat areas, a rise of just a few centimeters may result in a change in species dominance. This pattern is well developed in marshes and wet meadows.

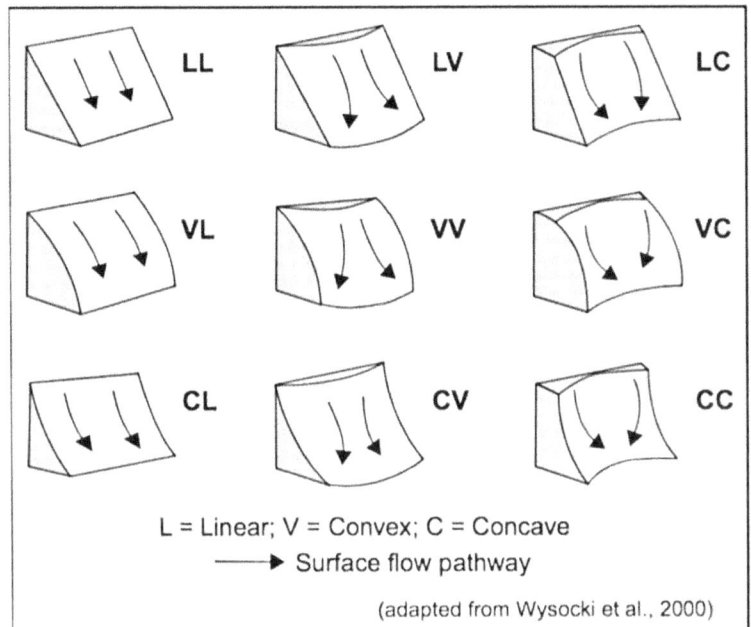

Figure 6: Slope Shape.
(From USDA Soil Survey Manual)

When a tree falls on a hill, the root mass raises the soil, leaving a depression. Over a period of time, the soil falls to the earth beside the depression, forming a mound. This surface feature, referred to as pit and mound topography, can play a significant role in seedling germination and survival. The feature is even more pronounced in undisturbed wooded swamp habitats, where it is often referred to as hummock and hollow topography. Because of shallow rooting, treefalls readily occur, leaving hummocks and hollows. Hollows often become flooded, while hummocks may remain above water level. These microfeatures provide a variety of microhabitats with varying degrees of moisture. Of course, the wind fallen trees themselves provide a highly organic environment that can last for a number of years. Many types of microfeatures may influence plant growth. For example, the presence of boulders, cobble stones, fallen logs, fallen branches, tree roots, tree stumps, dead shrub bases, live shrub bases, and proximity to such features are candidate features to evaluate (Jordan, & Hartman, 1995).

Exercise #32: Inclines

Describe slope characteristics within the area of interest, including slope gradient, slope class (see Table 10), and position and complexity of the slope (see Figures 5 and 6). When making a slope measurement with a clinometer, sight on an object directly up or down the slope. Find a point on the object that is at a height equivalent to your eye level. On long runs, pick objects as far up or down the slope as is practical to even out microrelief and microfeatures. Identify microfeatures such as pit and mound topography.

Depressions

These features occur as relatively sunken parts of the earth's surface and are of special ecological importance in low-lying areas surrounded by higher ground. If no natural surface flow outlet exists, the depression is said to be closed; a depression with a natural surface drainage outlet is open. Of course, almost all depressions will overflow if enough rainfall occurs, and so the outlet region needs to be examined. If flowing water can be seen draining the basin, then it is open. Under normal weather conditions, when no drainage water can be seen, examine the outlet region for a scoured channel. If present, the depression is considered open.

Depressions receive accumulations of eroded material, and therefore differ in their soil properties compared to slope sites. Depending on soil characteristics, surface flows, and depth to groundwater, depressions may be subject to ponding or high groundwater. In these conditions, soils may become anaerobic with relatively slow organic matter decomposition rates, resulting in a buildup of soil organic material.

Exercise #33: Depressions

Note and record depressions and any zonation patterns of vegetation. These observations may overlap with Exercise #31 Flats.

Geology and Soils

Geological processes have resulted in most of the major landscape and landform features observable today. Plate tectonics, bedrock formation, erosion, sediment transport, and sediment deposition impart a mosaic of surface features that affect plant growth. A few features have been chosen to illustrate the points discussed below.

Soils occur on the upper surface of land features. Over time, the near-surface materials that support plant life develop into various soil types because of the formation factors of relief, climate, parent material, weathering action, and biological processes that have or do exist at a particular site.

Geology

Volcanism, plate subduction, plate collision, and plate divergence produce mountain ranges, rift valleys, and major lakes across the land surface. In turn, these features have extensive effects on weather and water drainage. For example, on the west coast, prevailing winds move air masses ladened with moisture from the Pacific Ocean east across the Sierra Nevada Mountains. Precipitation may result as the air rises over these mountains, resulting in a relatively wet windward environment. On the east side, the mountains cast a "rain shadow," resulting in drier conditions. Another example is when a continental plate thins and moves apart. This may result in a continental rift zone, such as that found in the Great Basin (mostly in Nevada). The process led to the formation of horsts (raised, elongated blocks of the earth's crust) and grabens (elongated blocks displaced downward relative to horsts).

Bedrock

Rocks found in their original formations are referred to as bedrock. Exposed portions of bedrock make up relatively little of the earth's surface (probably less than 5 percent of the land surface, Hunt, 1984). Rocks are more frequently found as coarse fragments (small pebbles, cobblestones, boulders, etc.). Soil characteristics are greatly influenced by the parent material, which may be the local bedrock or material transported to a site. Coarse rock fragments found in transported material may indicate the original source of the parent material.

Rocks are classified by their methods of formation. Igneous rocks are formed by crystallization from a melt. Metamorphic rocks form by the recrystallization of former rocks under high pressure and/or temperature. Sedimentary rocks are formed by the consolidation of materials transported and deposited at a site. Specific types of rocks are identified by the types, amounts, and sizes of mineral grains of which they are composed. The type of rock determines how it decomposes through weathering. In turn, this weathering process affects soil nutrition. As an example, igneous rocks of high silica content and composed of large crystals weather slowly (felsic rocks), thus offering few nutrients. Weathering occurs more rapidly with fine-grained rocks containing mostly dark-colored minerals (mafic rocks), resulting in higher amounts of available nutrients in the soil.

Bedrock may be the causal agent behind local landforms, which in turn produce characteristics significant to plant habitats. A shallow depth to bedrock may restrict root growth. It can form a hydraulic restriction, resulting in saturated soil above it. Bedrock, by way of its shape and cracks, can direct groundwater towards various locations, including springs and seeps.

Exposed bedrock offers little in the way of a rooting medium, except if cracks are present that are large enough for root penetration. These cracks collect soil particles and small bits of organic matter, as well as water. Plants (chasmophytes) are able to make use of these resources, and both herbaceous and very slowly-growing woody plants may be found rooted in cracks.

Exercise #34: Bedrock

Note and record the presence of exposed bedrock and the type of rock (see Table 1 for geology map sources). Note and record areas of shallow bedrock (look for shallow root systems and very dry areas; probe with a knife or spade if shallow bedrock is suspected).

Unconsolidated Land Features

Weathering and rock and sediment transport have sculpted most of the land surface to its present configuration. Weathering turns bedrock into smaller particles and results from both mechanical and chemical processes. Erosion, by wind, water, glaciers, or gravity, transports rock and sediment and contributes to further weathering. Sediments eventually reach locations where energy from transport mechanisms is so reduced that deposition results. These unconsolidated materials cover the vast majority of land surfaces and are the medium in which soil develops. Surficial geology maps illustrate the locations and types of surficial features (see Table 1 for some map sources). Based on these maps, the local geologic setting and landscape position can be noted for an inspection-level outing.

A few landscape or landform features composed of unconsolidated sediments are described below. Features are arranged by the primary sediment transport mechanism.

Wind may transport small particles (sand, silt, and clay) and form loess deposits. In poorly vegetated areas, wind may produce sand dunes. Sand dune deposits may occur in recognizable landforms (e.g., parabolic, longitudinal, barchan, and transverse dunes). In deserts, wind may erode surface features, where the smaller particles are removed and the larger sands and pebbles are left behind.

Water-borne sediment may be deposited in marine, lacustrine, fluvial, and alluvial environments. Alluvial deposits are of primary interest here since the terrestrial environment is emphasized.

Streams and rivers tend to overflow their banks every one or two years when not contained by artificial means. In low-lying areas adjacent to the stream, the moving water transports sediments into an adjacent, nearly level floodplain and deposits the pebbles, sand, silt, and clay in this low energy environment.

As overbanking occurs, water velocity decreases, and some sediments (coarser materials, such as pebbles, gravel, and sand) are immediately deposited adjacent to the stream. These coarser materials may entrap smaller particles (sand, silt, and clay). The net result is that elevation increases immediately landward of the bank, forming a natural levee or embankment. The levee will slope downward both towards the bank and, on its opposite side, towards the floodplain. Of course, levees can be artificial, and in some cases, artificial levees are not immediately apparent. For example, to "improve" drainage, a stream may be straightened by excavation and the spoils placed directly along the new bank. Often, artificial levees along small streams can be detected because the stream channel has been straightened and the expected sinuosity is absent. Whatever the cause, the characteristics of a levee (e.g., increased soil drainage in the embankment) may lead to the presence of different vegetation compared to the adjacent floodplain.

Besides levees, other floodplain landforms can result from erosion or deposition. A stream may break through its natural levee and deposit sediments in a fan-like or other outspread shape, forming a floodplain splay. Marshy or swampy depressed areas, called backswamps, occur between natural levees and valley sides or floodplain terraces. One or more floodplain terraces may be created as a stream erodes downward and creates a new floodplain at a lower elevation. Terraces are step-like and characterized by flat summits (tread) and steeper descending slopes (riser).

When streams discharge into lakes or the sea, water velocity declines and sediment is deposited. Such areas are called deltas, and usually occur as a fan-shaped landform. Within a delta, the discharging stream channel may take a variety of shapes.

Pleistocene glaciation produced a host of conditions ranging from hills, slopes, depressions, and lakes, all of which are important to microclimate and soil formation. Additionally, the glaciers left a depositional mosaic of unconsolidated materials with differing textures.

Glaciers transported mineral material (clay, silt, sand, gravel, and boulders) and deposited them directly onto the land surface, or they released them into running meltwater, which then deposited the sediments. Drift is the general term for glacier-induced deposits.

Till is a kind of drift that is unsorted (a mixture of various-sized particles) and unstratified (no layering). It is deposited without subsequent reworking by meltwater. Moraines, composed of till, are mounds, ridges, or other topographically distinct accumulations of till. Deposits worked by meltwater tend to sort materials by size. These meltwaters can produce stratified drift and can form outwash plains, eskers, and kames.

A hill may be formed when outwash deposits fill a hole in the glacier. The hill, termed a kame, is left when the ice melts away. Glacial meltwater streams can flow within channels on top of glaciers and also through or under glaciers within tunnels. Ridges, called eskers, are deposited within these structures when sand and gravel are left behind and the ice melts. Blocks of ice may be left within glacial deposits that, on melting, leave a basin called a kettle. A kettle may contain a kettle pond.

Exercise #35: Unconsolidated Land Features
Identify and record the type of unconsolidated material within the area of interest using resources identified in Table

1, Chapter 2.

Soils

Plants depend on soil for their water and nutrient supplies and for anchorage. Very roughly, roots make up about 20 percent of a plant's body (Perry, 1989). However, observations of roots and the root environment are difficult at best. Soil composition, nutrient availability, and water content affect root growth, depth of rooting, and root form.

Soils can develop very slowly from bedrock, and develop more quickly and to a greater depth in unconsolidated surficial deposits. They form in the upper portions of surficial materials and extend to the limits of weathering and biological activity. Soil genesis is dependent on climate, the type of parent material available, weathering, soil biota (roots, burrowing animals, invertebrates, bacteria, fungi, and algae), and relief, and occurs over time.

Broadly speaking, two types of soil exist: mineral and organic soils. The A and B horizons of mineral soils often reach a depth of a few feet. They are made up of four components: mineral particles, water, air, and organic matter. Soil texture, an important characteristic of the mineral component, is based on the proportions of small particles present, i.e., sand, silt, and clay. Organic content in a mineral soil generally ranges below 12-18 percent by dry weight, and conversely, more than this would qualify as an organic soil. Air is found between soil particles in the pore space, as is water.

Organic soils are found in wetlands, including salt marshes, freshwater marshes, and wooded swamps. These soils are built up from the deposition of dead plant and animal materials. Organic matter can extend downward for only a few inches or up to 30 or more feet, although growing roots primarily occur only near the surface. These soils, or portions of them, may be composed of peat, mucky peat, or muck, depending on how much decomposition has occurred. Peat is composed of largely undecomposed organic matter; muck is made up of well-decomposed organic soil material; and mucky peat has an intermediate degree of decomposition.

Organic matter plays a major role in mineral soils. Duff refers to the surface organic layer and consists of fallen plant material. It includes the surface litter and the underlying pure humus. Humus is the well-decomposed organic matter in mineral soils and is capable of being transported as small particles and dissolved compounds downward as water infiltrates into the mineral soil.

Soil texture (relative proportions of sand, silt, and clay) affects a plant's ability to obtain water, oxygen, and nutrients. Sand consists of mineral particles ranging from 0.05 to 2.0 millimeters in diameter. Individual silt particles range from 0.002 to 0.05 millimeters, and silt feels like talcum powder. Clay particles are less than 0.002 millimeters in diameter.

Clay is weathered from larger particles and exhibits specific properties affecting nutrient and water availability. The structure of clay is such that its surface has numerous negative charges available for ion attachment. A unit volume of clay contains an extremely large surface area because of the particles' small size. Clay surfaces attract and hold on to positively charged ions (cations). Most of the macro- and micronutrients obtained from the soil are cations, which plants can obtain from clay through a process of cation exchange.

Soils are placed into textural classes, with each class having different characteristics. Figure 7 illustrates the range of sand, silt, and clay found within each class. The twelve classes, listed in order of increasing proportion of fine particles, are:
- sand
- loamy sand
- sandy loam

- loam
- silt loam
- silt
- sandy clay loam
- clay loam
- silty clay loam
- sandy clay
- silty clay, and
- clay.

The sand, loamy sand, and sandy loam classes may be further divided into coarse, fine, or very fine sand. Coarser soils cannot hold as much water as finer soils; hence coarse-textured soils tend to be more drought-prone. (On the other hand, fine soils like clays can hold tightly to water, making it difficult for plants to obtain.) Finer textures have greater total particle surface areas, potentially leading to greater nutrient holding capacity.

An experienced observer can estimate texture class based on qualitative criteria. Criteria include how the soil feels (gritty, smooth, or sticky) and how it behaves when rubbed between fingers to form a ribbon. Individual grains of sand can be seen without magnification and feel gritty. Individual silt particles can not be seen without magnification; silt feels smooth when dry or wet. Clay soils are sticky (this can vary depending on geographic location and parent material). Figure 8 illustrates the texture estimation process used for field determinations.

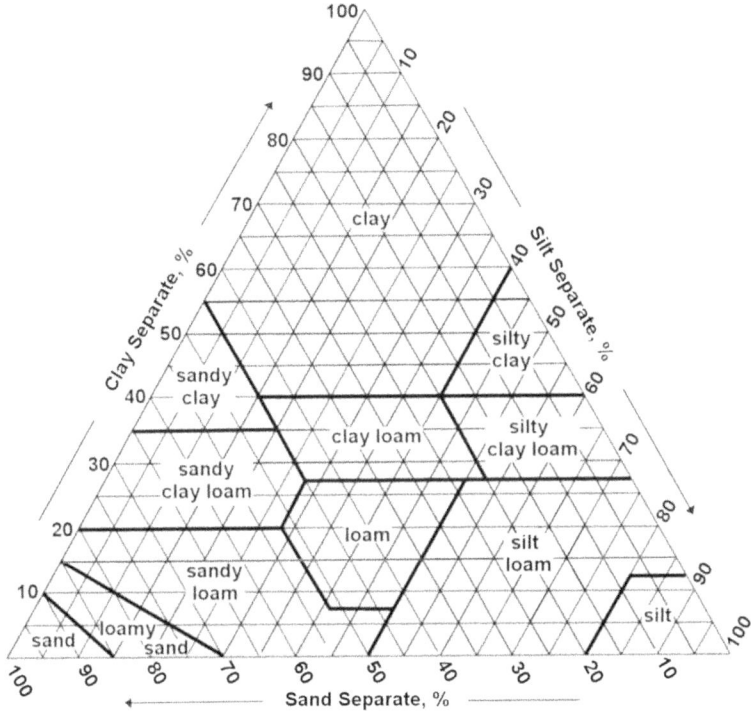

Figure 7: USDA Soil Texture Triangle.

Soil color is used in the classification of soil types. It is determined for each horizon and horizon component (e.g., matrix, concretions, nodules, zones of depletion, redoximorphic features, etc.).

While color may be evaluated when soil is either dry or wet, the exact color will differ with wetness, and it is recommended here that moist soil be used. Apply enough water to wet a sample without causing the surface to glisten.

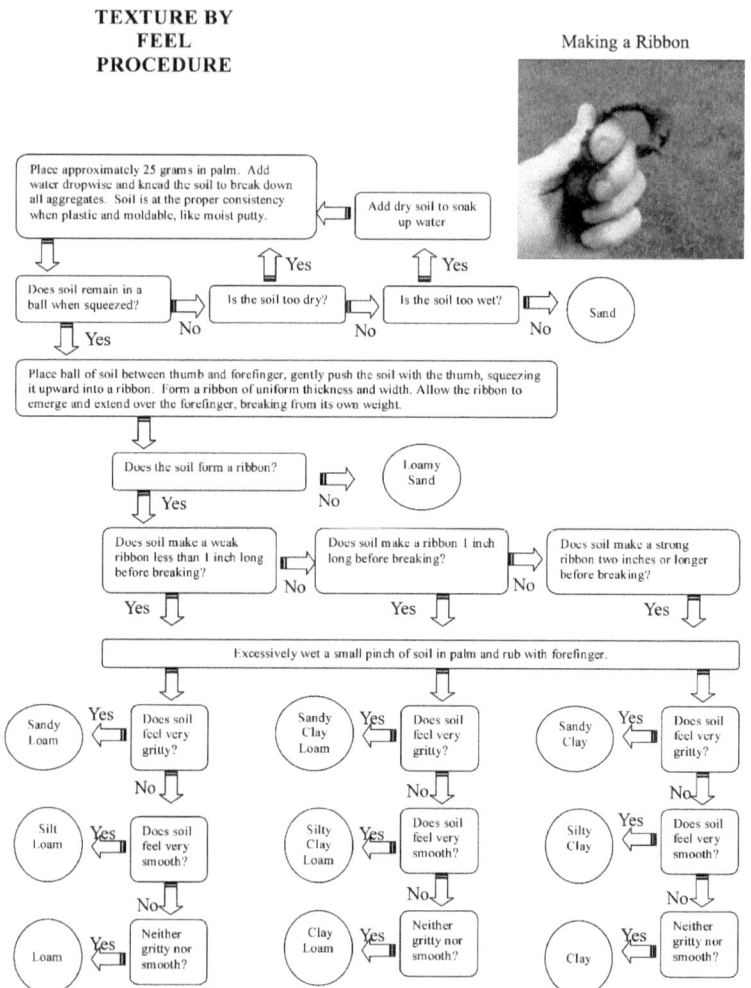

Figure 8: Texture by Feel Procedure.
(From USDA: https://www.nrcs.usda.gov/sites/default/files/2022-10/texture_feel.pdf.)

If the soil sample is already wet, wait a few minutes until the surface no longer glistens. Using Munsell soil color charts, determine hue, value, and chroma by placing a sample behind the holes in the charts and matching soil color with a color chip. Specific instructions are normally included in a Munsell Soil Color book.

Hue refers to the specific color shown on a particular soil chart. Chips on each chart vary in **value** from light to dark. Chips also vary in **chroma**, which is the chip's intensity (relative purity or strength of the spectral color). Soil color is discussed in detail in the Soil Survey Manual (Soil Science Division Staff, 2017).

Hydric (wetland) soils are of particular note since they indicate the presence of various types of stressors on plants. Wetlands occur in areas that are inundated, or have soil saturation near the soil surface, for at least a few days during the growing season. Such water saturation or inundation leads to anaerobic conditions, which retard decomposition processes, leading to a buildup of organic matter that can be many feet thick. Anoxic conditions also lead to reduced forms of a number of elements found in the soil. For example, sulfates may be reduced to hydrogen sulfide, producing a distinct "rotten egg" odor when the soil is disturbed.

Soil microbes reduce iron from the ferric form (rust-colored) to the ferrous form and manganese from the manganic to the manganous form. These reduced forms are capable of being translocated from and accumulated into soil layers by water moving through the soil. This process leads to distinct color changes within the soil. In general, hydric soils with a layer of dark organic material 20 cm (8 inches) to 40 cm (16 inches) thick will be underlain by one or more depleted horizons. These depleted horizons will be light in color (value ≥ 3) with a chroma of ≤ 2. But many variations of horizon position, thickness, and color exist that can be used to indicate a hydric soil.

Examine small roots to determine the presence or absence of oxidized rhizospheres. In areas with fluctuating water tables, the soil surrounding rootlets may be rust-colored. This results from the leakage of oxygen from roots during anaerobic periods, which in turn leads to the deposition of ferric compounds.

Wetland areas may not have standing water when observed, and some may not even be inundated during the growing season. All wetlands will at least have saturated soil near the ground surface for a period of time when plants are active. When examining soils for hydric indicators, dig a hole that is deep enough to expose hydric soil characteristics [usually about 20 inches (50 cm) from the soil surface]. A list of hydric soil field characteristics is available in "Field Indicators of Hydric Soils in the United States" (United States Department of Agriculture, Natural Resources Conservation Service, 2018). These indicators are best used near the edge of a wetland, and any particular wetland soil may not exhibit all characteristics.

Some common characteristics include:

- A deep [typically 40 cm (16 inches) or more] of muck, mucky peat, or peat extending from the upper surface (a histosol, or histel in areas with permafrost)

- Mineral soils that have a surface layer of organic matter extending 20 cm (8 inches) or more and are underlain by a mineral soil with a chroma of 2 or less

- Sulfidic material that produces a "rotten egg" when the test hole is first opened

- Soils that are neutral gray, greenish, or bluish gray (i.e., gleyed soil) near the surface

- Soils with a matrix chroma of 0 or 1 and values of 4 or higher near the bottom of the O horizon (see "**Soil Profile**" below for a discussion of horizons)

Soil Profiles

As soils develop over time, soil building processes form horizontal layers (horizons) near the land's surface. The combination of horizons is referred to as a "soil profile", and profiles form the basis for classifying soils.

The uppermost soil layer is the O horizon, composed of slightly to well-decayed plant material. (Fresh organic deposits are not included.) The first mineral layer is the A horizon. It has an accumulation of humified organic matter mixed in with the mineral particles and is generally dark in color. While water percolation brings organic matter and organic compounds downward and enriches the A horizon, it also moves dissolved minerals and clay downward out of the A horizon. Sometimes a layer develops below the A horizon, referred to as the E horizon, that has lost silicate clay, iron, aluminum, or some combination of these to a lower horizon. The E horizon generally has a light color. The B horizon is found next. In those soils without an E horizon (which is common), the B layer is found under A. "The B horizon has distinctive characteristics, such as (1) accumulation of clay, sesquioxides, humus, or a combination of these; (2) prismatic or blocky structure; (3) redder or browner colors than those in the A horizon; or (4) a combination of these" (Soil Survey Staff, Web Soil Survey glossary). The C horizon occurs next. It is a mineral layer that has not been substantially affected by soil-forming processes. A particular soil type may be missing a horizon, or additional horizons not considered here may be present. For a complete explanation of layering and layer modifiers, see the Soil Survey Manual (Soil Science Division Staff, 2017).

Exercise #36: Soil Profile, Texture, and Color

Dig a soil pit to a depth of about two feet or until refusal. Examine, identify, and measure the depth and thickness of the major layers of the soil (i.e., O, A, E, and B) and the depth to the C horizon. Estimate the soil texture and color for each layer. (Note: Profile measurements begin below any fresh leaf or needle fall.)

Web Soil Survey

Many site and soil characteristics can be obtained from the Web Soil Survey (Soil Survey Staff). For an inspection-level trip, it is beneficial to have this information at hand. The first step in using the Web Soil Survey is to identify an area of interest. Then, by clicking on the soil map tab, a map of soil types will appear along with a list of map units, including soil types present in the area. Clicking on a soil type will bring up a report containing the information under consideration here. As an example, Table 12 illustrates a report generated for an area on Cape Cod in Massachusetts. All reports follow the same general outline. This report is used to discuss line items found in reports.

Map Unit Setting and Composition: Data found under these headings indicate a range of conditions where the map unit is found, not just within the Area of Interest. Note that about 80% of the unit is composed of Carver coarse sand and that about 20% of the unit is made up of other types of soils. Keep this in mind when on an inspection-level outing, since the particular site observations may lie in one of the soil types not mapped.

Table 12: Sample Report from the Web Soil Survey
Barnstable County, Massachusetts

252B—Carver coarse sand, 3 to 8 percent slopes

Map Unit Setting
National map unit symbol: 2y07x
Elevation: 0 to 240 feet
Mean annual precipitation: 36 to 71 inches
Mean annual air temperature: 39 to 55 degrees F
Frost-free period: 140 to 240 days
Farmland classification: Not prime farmland

Map Unit Composition
Carver, coarse sand, and similar soils: 80 percent
Minor components: 20 percent
Estimates are based on observations, descriptions, and transects of the mapunit.

Description of Carver, Coarse Sand

Setting
Landform: Moraines, outwash plains
Landform position (two-dimensional): Summit, shoulder, backslope, footslope, toeslope
Landform position (three-dimensional): Crest, head slope, nose slope, side slope, tread
Down-slope shape: Convex, linear
Across-slope shape: Linear
Parent material: Sandy glaciofluvial deposits

Typical profile
Oi - 0 to 2 inches: slightly decomposed plant material
Oe - 2 to 3 inches: moderately decomposed plant material
A - 3 to 7 inches: coarse sand
E - 7 to 10 inches: coarse sand
Bw1 - 10 to 15 inches: coarse sand
Bw2 - 15 to 28 inches: coarse sand
BC - 28 to 32 inches: coarse sand
C - 32 to 67 inches: coarse sand

Properties and qualities
Slope: 3 to 8 percent
Depth to restrictive feature: More than 80 inches
Drainage class: Excessively drained
Runoff class: Low
Capacity of the most limiting layer to transmit water (Ksat): Moderately high to very high (1.42 to 14.17 in/hr)
Depth to water table: More than 80 inches
Frequency of flooding: None
Frequency of ponding: None
Maximum salinity: Nonsaline (0.0 to 1.9 mmhos/cm)
Available water supply, 0 to 60 inches: Low (about 4.3 inches)

Table 12 (Continued)

Interpretive groups
    Land capability classification (irrigated): None specified
    Land capability classification (nonirrigated): 3s
    Hydrologic Soil Group: A
    Ecological site: F149BY005MA - Dry Outwash
    Hydric soil rating: No

Minor Components

Deerfield
    Percent of map unit: 10 percent
    Landform: Outwash terraces, outwash plains, kame terraces, outwash deltas
    Landform position (three-dimensional): Tread
    Down-slope shape: Linear
    Across-slope shape: Concave
    Hydric soil rating: No

Hinckley
    Percent of map unit: 5 percent
    Landform: Moraines, eskers, kames, outwash deltas, outwash terraces, outwash plains, kame terraces
    Landform position (two-dimensional): Summit, toeslope, shoulder, backslope, footslope
    Landform position (three-dimensional): Side slope, crest, head slope, nose slope, riser, tread
    Down-slope shape: Convex
    Across-slope shape: Convex
    Hydric soil rating: No

Merrimac
    Percent of map unit: 3 percent
    Landform: Kame terraces, outwash deltas, outwash terraces
    Landform position (three-dimensional): Riser, tread
    Down-slope shape: Linear
    Across-slope shape: Linear
    Hydric soil rating: No

Mashpee
    Percent of map unit: 2 percent
    Landform: Depressions, drainageways, terraces
    Landform position (three-dimensional): Tread
    Down-slope shape: Concave
    Across-slope shape: Concave
    Hydric soil rating: Yes

Data Source Information

Soil Survey Area: Barnstable County, Massachusetts
Survey Area Data: Version 19, Sep 9, 2022

More information may be found by searching for the soil series, including a distribution map, at https://soilseries.sc.egov.usda.gov/. The Web Soil Survey glossary defines a series as "[a] group of soils that have profiles that are almost alike, except for differences in texture of the surface layer. All the soils of a series have horizons that are similar in composition, thickness, and arrangement." Hence, searching under Carver (the series name) will bring up access to this information.

Setting: As shown in Table 12, the <u>Landform</u> is identified first, which indicates that this soil type is found on moraines and outwash plains. Various landform definitions may be found in "Title 430 – National Soil Survey Handbook Part 629 – Glossary of Landform and Geologic Terms" (U.S. Department of Agriculture).

<u>Landform position (two-dimensional)</u> is described earlier (see Figure 5). <u>Landform position (three-dimensional)</u> refers to the locations within a landscape where the soil type is found. Because of the variety of landscapes across the nation, specific terminology has been developed for four general landscape types: hills, terraces and stepped landforms, mountains, and flat plains. See the Soil Survey Manual (Soil Science Division Staff, 2017) for a discussion of three-dimensional landform positions.

The setting's <u>shape (down-slope and across-slope)</u> is described earlier (See Figure 6).

Typical Profile: The location of each horizon within the profile is identified in this section of the report. Capitol letters identify major horizons. Sometimes a single major horizon may be subdivided based on observable characteristics. In this case, the capitol letter is followed by a number. Certain characteristics of a major horizon may be identified by a suffix. For example, the sample profile above identifies a horizon labeled Bw1. This horizon is the first sub-layer found in the B horizon. The Soil Survey Manual

(Soil Science Division Staff. 2017) states that the w "symbol is used only with B horizons to indicate the development of color or structure, or both, with little or no apparent illuvial accumulation of material. A complete description of horizon designations may be found in the Soil Survey Manual.

Properties and Qualities: Slope has been discussed above under **Inclines**. Depth to restrictive feature identifies the position of any feature that may restrict root growth.

Natural drainage class refers to the frequency and duration of wet periods. Seven classes of natural soil drainage are recognized:

- excessively drained,
- somewhat excessively drained,
- well drained,
- moderately well drained,
- somewhat poorly drained,
- poorly drained, and
- very poorly drained.

Exact descriptions of each drainage class, taken from the Soil Survey Manual (Soil Science Division Staff, 2017), are presented in Appendix 1. Drainage classes describe the rate at which free water (i.e., water that can drain out by gravitational force) is removed. Wetland soils typically have very poorly or poorly drained soils.

Runoff refers to the water that flows off the surface of the land without sinking into the soil. The Runoff class identifies one of four classes, ranging from low to high, to which the soil type belongs.

The Capacity of the most limiting layer to transmit water refers to downward water movement in the soil. As water moves down through a soil, it meets resistance from horizons. A particular soil layer may pose greater resistance to water transmission than others in the profile, thus controlling soil drainage. If severe, the most limiting layer can create prolonged saturation above it. If this saturation occurs in the rooting zone, plant growth and species composition will be affected.

Depth to water refers to the position below which the soil is saturated with water. This position may vary over the seasons. Many plants are able to grow tap roots or sinker roots, which may reach the water table. Usually roots are confined above the water because of the low oxygen content within the saturated zone. If the water table reaches near or to the surface, wetland conditions prevail. This results in a buildup of organic materials since decomposition processes are slow. Wetland soils include peats, mucks, and mucky mineral soils. In order to live in wetlands, plants must be adapted to the low oxygen environment and associated soil chemistry.

Wetland plant communities are primarily composed of facultative, facultative wet, and obligate species. Mineral soils with a depth to the seasonally high water table exceeding 18 inches are usually in upland conditions that support upland, facultative upland, and facultative species. The depth of the water table is an obvious characteristic that will affect plant distribution and abundance.

Prolonged flooding or ponding produces major effects on vegetation. Ponding is differentiated from flooding in that it is standing water on soil in closed depressions. It can only be removed by percolation or evapotranspiration. Flooding is removed as the flood waters recede.

Salinity refers to the amount of soluble salts in soil or water. It may be estimated by measuring soil conductivity, which is reported in mmhos/cm.

Water is located within soils in pore spaces of varying sizes and is also adsorbed to soil particle surfaces. After saturation by rainfall, excess water (called gravitational or free water) drains out of the soil. At this point, the soil is at field capacity. Without replenishment, the remaining soil water declines due to evapotranspiration. A point is reached where water becomes unavailable to plants (the wilting point) because it is too tightly held by adhesion to soil particles. The amount of water held by soil between the wilting point and field capacity is referred to as <u>available water capacity</u>. This capacity, in inches, in a 60-inch profile or to a limiting layer is expressed as:

- very low: 0 to 3,
- low: 3 to 6,
- Moderate: 6 to 9,
- high: 9 to 12, and
- very high: more than 12

Exercise #37: Web Soil Survey Data

Using the web soil survey, locate your areas of interest and print off soil maps. Click on each map unit and print off its description.

Soil pH, Nutrition, and Other Soil Parameters

Soil pH influences a number of characteristics that affect plant growth and hence, the presence and abundance of a species. It regulates the availability of plant nutrients, the activity and type of soil microorganisms, and soil toxicity. It also affects the solubility of soil minerals and the rate of soil weathering.

The degree of soil acidity or alkalinity is measured by the pH of a 1:1 soil to water extract. Because H^+ ions are positively charged, they may be adsorbed to the negatively charged surfaces of soil particles such as clay and humus. When H^+ ions are released into the soil, they may displace cation nutrients stored on the surface of these particles.

H^+ ions in the soil originate from a variety of sources, including dissolved carbon dioxide, which forms carbonic acid in water. Carbon dioxide may originate from the soil atmosphere, decomposing organic matter, and root and microbial respiration. Roots release organic acids, and acid rain acidifies the soil. Organic acids may be released into the soil from decaying organic matter. Hydrogen sulfide from decaying organic matter or high sulfur mineral deposits can react with water to form sulfuric acid.

The pH of a soil is also referred to as its soil reaction. Reaction classes have been established, as shown in Figure 9. Soil reaction affects nutrient availability, with each nutrient having its own maximum availability depending on pH. The National Soil Survey Handbook (USDA NRCS) states:

> *Strongly acid or more acid soils have low extractable calcium and magnesium; a high solubility of aluminum, iron, and boron, and a low solubility of molybdenum. In addition, these soils may possibly have organic toxins and generally have a low availability of nitrogen and phosphorus. At the other extreme are alkaline soils. Calcium, magnesium, and molybdenum are abundant where there is little or no toxic aluminum and nitrogen is readily available. If pH is above 7.9, the soils may have an inadequate availability of iron, manganese, copper, zinc, and especially phosphorus and boron.*

A range in pH between 6 and 7.5 is often described as optimum for plants; however, this relates to common crop species.

The ecology of individual species can be affected by soil pH. Also, general community types can be expected under certain pH regimes, as shown in Figure 9.

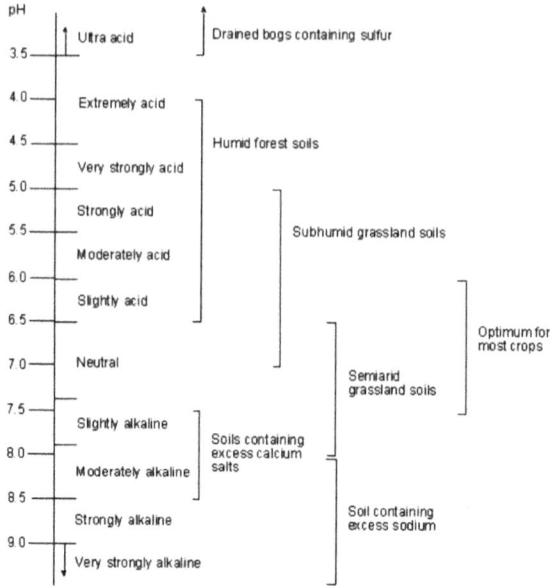

Figure 9: Soil pH, Ranges for pH classes, and Associated Soil Conditions.
(From: Soil Quality Test Kit Guide, USDA NRCS.)

Elements necessary for plant life are categorized as either macronutrients (those needed in relatively large amounts) or micronutrients (needed in small amounts). The macronutrients consist of nitrogen, phosphorus, potassium, calcium, sulfur, magnesium, carbon, oxygen, and hydrogen. Carbon and oxygen can be obtained from atmospheric oxygen and carbon dioxide. Water provides a source of hydrogen and oxygen, and of course both water and carbon dioxide are key in the photosynthetic process, which produces high energy carbon compounds and free oxygen. The remaining macronutrients, as well as micronutrients, are primarily obtained from soil.

Nutrients play important roles in the successful growth and distribution of plants, either because availability is inadequate or overly abundant. Chief among those macronutrients that often occur in insufficient quantities are nitrogen and phosphorus. On the other hand, some necessary nutrients, such as sulfur, can be a detriment when too much is present. The rotten egg odor of many wetlands, particularly when soils are disturbed, results from the presence of hydrogen sulfide. In high amounts, this compound can be toxic to plants.

Micronutrients consist of iron, boron, chlorine, manganese, zinc, copper, molybdenum, and nickel. They are needed only in trace amounts, which are generally available from the soil. Nevertheless, in certain environments, micronutrients may limit growth either because one or more are in short supply or are overly abundant. For example, plants suffer chlorosis (yellowing of leaves) if inadequate iron is available. Many plants adjacent to the coast are negatively impacted by the abundance of sodium chloride in salt spray. Some soils in deserts may contain high levels of sodium, thus affecting species presence.

As mentioned earlier, nitrogen and phosphorous are often of limited availability in ecosystems. The largest reservoir of soil nitrogen resides in soil organic matter. It is released into the soil water as decomposition occurs, where it can be absorbed by plant roots. Mycorrhizae allow plants to harvest nitrogen from fungal associates as the fungi absorb it from soil organic matter. Nitrogen-fixing root nodules provide a source of nitrogen for a number of plants. Dissolved nitrogen anions are easily lost from an ecosystem when water transports them out of the soil. Phosphorous originates from decaying organic matter and from minerals in the soil. It is taken up by plant roots from the soil water. However, phosphate compounds have low solubility in water, so even if the soil mineral fraction is high in phosphorus, it may not be available. The availability of both elements may be reduced by either high or low pH conditions.

The majority of macro- and micronutrients are present as cations. These positively charged ions attach to the negatively charged clay surfaces and form a reservoir of potentially available nutrients. Attached cations can be released for absorption through the process of cation exchange. Hydrogen ions are produced from the release of organic acids and carbon dioxide into the soil water. The hydrogen ions exchange places with cations on the clay surface, releasing them for root absorption.

The clay fraction of a soil provides the largest mineral surface area and is more active in cation-exchange than silt or sand. Organic matter has an exceedingly high cation-exchange capacity, and also has nutrients locked up in organic compounds. Cation-exchange capacity refers to the total capacity of a soil to hold exchangeable cations. Other things being equal, soil types with higher fractions of clay tend to hold more cations. Sandy soils have a relatively low capacity.

Many types of soil parameters may be useful when evaluating and comparing areas of interest. Table 13 lists several of these that are readily available from the web soil survey (Soil Survey Staff).

Table 13: Useful soil parameters

Soil Parameter	Comments
Cation-exchange capacity (milliequivalents per 100 grams)	Total amount of extractable cations that can be held by the soil. Lower values suggest fewer nutrient cations available compared to higher values.
Effective cation-exchange capacity (milliequivalents per 100 grams)	Refers to the sum of extractable cations plus aluminum and is determined for soils that have a pH of less than 5.5. Lower values suggest fewer nutrient cations available compared to higher values.
Electrical conductivity (decisiemens per meter)	Electrical conductivity is a measure of the concentration of water-soluble salts in soils. It is used to indicate saline soils.
Calcium carbonate (%)	The percent of carbonates, by weight, in the fraction of the soil less than 2 millimeters in size. The availability of plant nutrients is influenced by the amount of carbonate

Soil Parameter	Comments
	in the soil.
pH	A measure of acidity or alkalinity. It regulates the availability of plant nutrients, the activity and type of soil microorganisms, and soil toxicity. It also affects the solubility of soil minerals and the rate of soil weathering.
Sodium adsorption ratio	The amount of sodium relative to calcium and magnesium. Soils with values of 13 or more may be characterized by an increased dispersion of organic matter and clay particles, reduced saturated hydraulic conductivity (Ksat) and aeration, and a general degradation of soil structure.
Organic matter (%)	Estimates the percent of organic matter, by weight, in the fraction of the soil less than 2 millimeters in size. It may increase available water capacity, water infiltration, and soil organism activity. It is a source of nitrogen and other nutrients for plants and soil organisms.

Exercise #38: Soil Parameters From The Web

Soil parameters listed in Table 12 may be determined by examining the Web Soil Survey (Soil Survey Staff). After identifying an area of interest, click on the "Soil Data Explorer" tab, followed by the "Soil Properties and Qualities" tab. Under "Properties and Qualities Ratings," choose "Soil Chemical Properties," which lists all parameters found in Table 12 with the exception of organic matter. Organic matter can be found listed under "Soil Physical Properties".

Choose a parameter, and, under advanced options, choose "Depth Range (Weighted Average)". Choose the top and bottom depths for analysis (rooting zone would be appropriate, e.g., 0 to 12 inches). Click on "View Rating" to see parameter values for each soil in the area of interest. Prepare a table of values for all parameters for comparison between areas of interest.

Flooding and High Groundwater

Temporary, seasonal, or permanent flooding locations, and high groundwater are key environmental characteristics. As examples, inundation may result in the water dispersal of seeds and propagules. Exposure to wave action may disrupt substrates and influence species presence along ponds, lakes and coastal shorelines. Flooding stream flows may result in undermined and exposed root systems along and near banks. Tidal flooding is a driving force behind saltmarsh distributions. High groundwater levels influence the types of wetland plant communities that may develop.

During periods of rainfall or snow melt, water that is unable to infiltrate soil moves downhill as surface flow. Flows first occur just below the litter zone, and as rain continues, they coalesce into rivulets. Water flows perpendicular to contour lines, and hence topography plays an important role in the amount of water a particular area receives. Swales accumulate more water than adjacent slopes, often resulting in the growth of different plant species compared to neighboring areas.

Areas subject to flooding or ponding often support different species compared to adjacent environments. Timing, duration, and depth of surface water play major roles in the species composition of such areas. In general, plants are able to tolerate brief exposures to inundation (measured in hours or days) and even longer exposures if they are dormant. Special adaptations are required to survive long term (weeks and months) or permanent surface water, particularly during the growing season.

Permanent inundation may result in an aquatic ecosystem with submerged, floating-leafed, and emergent vegetation growing within certain water depth zones. Variations in the depth and duration of a water table near the ground surface also result in unique wetland communities. For example, swamps, marshes, wet meadows, bogs, and fens develop based on their particular hydrology.

The presence of nearby ephemeral or perennial surface water may increase plant growth and vigor, resulting in a strip of riparian vegetation along the waterbody.

In upland conditions after a significant rain event, saturated soils drain; air is able to replace water in soil pore spaces, thus providing a source of oxygen to plant roots. As roots and soil organisms respire, oxygen is used, but diffusion from the atmosphere replenishes lost oxygen. The diffusion rate into the soil drops under saturated conditions, and the oxygen supply becomes depleted. Oxygen depletion occurs not only below the water table but directly above it because capillary forces pull water upward, thus waterlogging this zone of soil. Plants living in waterlogged soils must be able to withstand adverse wetland conditions brought about by deoxygenation.

Groundwater is not static. It flows along hydraulic pressure gradients, which usually mean downhill, but gradients can also force groundwater in horizontal or vertical directions, resulting in springs and breakouts on slopes.

Evidence of Temporary Inundation

Stress from temporary inundation affects seed germination and survival, plant growth, and floristic composition. Substrates may be eroded and top soil lost due to moving water or wave action. Some areas receive new substrates and nutrients as sediments drop out in still or slow-moving locations. The degree of stress increases with the duration of the inundation.

As water recedes, it leaves a number of characteristics that can be used to identify locations of temporary inundation. Some possible indicators of past inundation (taken primarily from Bureau of Land Management, 2021 and Jackson et al., 2022) include:

- exposed shorelines of both lentic and lotic environments, bare of vegetation;
- wrack lines laid down during high tide;
- water marks on trees, boulders, etc. Marks include stained or silt-covered areas and changes in plant or lichen growth occurring at one elevation;
- water-stained leaves on the ground (usually flattened and dull gray or black in color);
- sediment deposits on plants, leaves, or the ground;
- drift lines (accumulations of plant material or debris) on the ground or in the branches of woody vegetation roughly parallel to nearby streams and rivers;
- scoured areas (relative absence of leaf litter and other debris on the ground, or where fine soils have been washed away, leaving gravel and cobble);
- drainage patterns left by flowing water (e.g., braided patterns in the sediments, channels in the leaf litter, vegetation bent in one direction by the force of running water);
- presence of fingernail clams and aquatic snails or their shells; and
- caddisfly cases (tubelike cases made of leaf fragments, twigs, pine needles, or sand).

Exercise #39: Indicators of Past Inundation

Inspect an Area of Interest for water bodies, waterways, and depressions. Identify the limits of past inundation using the above characteristics.

Riparian Environments

A zone of vegetation adjacent to lentic or lotic water bodies can often be seen that may be especially green, tall, and luxuriant. The species composition may or may not be different from areas further away. The vegetation may be dominated by either upland or wetland species, depending on climate, terrain, and soil conditions.

> *Riparian areas are closely associated with water and topographic relief; they are distinct from either wetland or upland. Riparian areas lack the amount or duration of water usually present in wetlands, yet their connection to surface or subsurface water distinguishes them from adjacent uplands.* (U.S. Fish and Wildlife, 2019)

Various definitions of "riparian" exist, but the definition adopted by the U.S. Fish and Wildlife Service is presented here:

> *Riparian areas are plant communities contiguous to and affected by surface and subsurface hydrologic features of perennial or intermittent lotic and lentic water bodies (rivers, streams, lakes, or drainage ways). Riparian areas have one or both of the following characteristics:*
>
> *1. Distinctly different vegetative species than adjacent areas.*
> *2. Species similar to adjacent areas but exhibiting more vigorous or robust growth forms.*
>
> *Riparian areas are usually transitional between wetland and upland.* (U.S. Fish and Wildlife, 2019)

The verdant nature of riparian vegetation is especially

characteristic along water bodies and waterways in the western United States in zones where evaporation exceeds annual precipitation. Using remote sensing, the U.S. Fish and Wildlife Service is currently mapping riparian zones where mean annual evaporation exceeds mean annual precipitation by 10 inches or more (see U.S. Fish and Wildlife, 2019).

Exercise #40: Riparian Zone Identification
Inspect an Area of Interest for water bodies or waterways. Identify verdant zones that meet the definition of riparian. Riparian zones in certain areas in the western states have been mapped by the U.S. Fish and Wildlife Service, and this information is available on the Wetlands Mapper site (see Chapter 2, Table 1).

Wetland Hydrology

The most prominent and defining feature of a wetland is the presence of a water table at or near the surface, or where the land is covered by shallow water. This simple statement embraces a multitude of conditions. For example, an area could be susceptible to inundation or high groundwater because it is in a flood plain, because it occupies a depression, because it is located on a slope where break-out occurs, or because of the influence on groundwater from a nearby waterbody. The source of water may be precipitation, overland flow, stream inputs, or groundwater. Loss of water from a wetland may result from recession of flood waters, overtopping of a basin, evapotranspiration, outflow by a stream, or groundwater drainage. While these conditions represent wetland inputs and outputs, there is also considerable variation in internal hydrology. A wetland may contain stream channels, groundwater may move horizontally through a wetland, or water levels may move up and down from precipitation and evapotranspiration. Many of the foregoing conditions may operate together to determine the hydrology of a wetland.

A number of classification systems exist that emphasize various wetland characteristics, such as the type of vegetation present or the type of hydrogeomorphic features they exhibit. Brinson (1993) offers a brief discussion of a number of systems based on hydrogeomorphology, and discusses an approach which emphasizes three wetland features: geomorphic setting, water sources, and hydrodynamics. "Geomorphic setting is the topographic location of the wetland within the surrounding landscape. The types of water sources can be simplified to three—precipitation, surface or near-surface flow, and groundwater discharge. Hydrodynamics refers to the direction of flow and strength of water movement within the wetland" (Brinson, 1993).

Hydrogeomorphic wetland classes are discussed by Smith et al. (1995) and the USDA Natural Resources Conservation Service (2008). These sources have been used in developing the information presented below. Seven major wetland classes are discussed:

- Riverine,
- Depressional,
- Slope,
- Mineral soil flats,
- Organic soil flats,
- Lacustrine fringe, and
- Estuarine Fringe.

<u>Riverine</u> wetlands are associated with stream channels that carry either intermittent or perennially moving water. They occur in flood plains and riparian corridors. Water is supplied to the wetland primarily by overbank and/or subsurface flows from the channel to wetlands. Other water sources may come from adjacent uplands. For example, interflow (water movement in the unsaturated soil zone during and immediately after a precipitation event) and overland flow may provide water. Precipitation also contributes water, as do tributaries, if present. At a stream's headwater, where channel morphology is absent, a riverine wetland may

transition into a slope or depressional wetland. Integration downstream into poorly drained flats or uplands may also occur.

As the name suggests, <u>depressional</u> wetlands are formed within topographic depressions. The major water sources are precipitation, groundwater discharge, and interflow and overland flow from adjacent uplands. The depression may be open or closed; however, if open, at least some of the topography lies below the outlet invert level. Thus, the direction of flow is typically toward the topographic lows, where surface water may accumulate. Vertical water fluctuations, primarily seasonal, are dominant. Water loss occurs from evapotranspiration and, if they are not receiving groundwater discharge, by infiltration into groundwater. Intermittent or perennial surface drainage may occur through an outlet.

Sloping ground, whether on steep hillsides or very shallow slopes, may lead to areas of groundwater discharge to the surface. Under these circumstances, <u>slope</u> wetlands may develop. Depressional storage is usually lacking. Ground water, precipitation, and interflow from surrounding uplands form sources of water. Downslope, unidirectional water flow is the principal norm. Methods of water loss include saturated subsurface and surface flows, and evapotranspiration. Stream channels, which serve only to convey water away from the wetland, may occur.

<u>Mineral soil flats</u> may occur in interfluves (high flat terrain located between watercourses), extensive relic lake bottoms, or large historic flood plain terraces. Their main water source is precipitation. They are distinguished from depressional and shallow slope wetlands because groundwater discharge is not a source of water. Vertical water fluctuations are dominant. Water is lost by evapotranspiration, overland flow, and infiltration. They differ from flat upland areas by having poor vertical and low lateral drainage. Peat can accumulate, eventually developing an organic soil flat.

Extensive peatlands may be found as organic soil flats. Their elevation and topography are controlled by the vertical accretion of organic matter. Similar to mineral soil flats, organic flats may occur on flat interfluves. They may also be located in areas where depressions have become filled with peat to form relatively large flat surfaces. Precipitation is the dominant source of water. Water loss may occur due to saturated overland flow and infiltration. Raised bogs share a number of these characteristics, but are a separate class because of their form (convex) and edaphic conditions.

Lacustrine fringe wetlands occur adjacent to lakes and have their water table maintained by the lake water elevations. These wetlands may include floating mats attached to the land. Precipitation is an additional source of water. Groundwater discharge may dominate where the wetland intergrades with uplands or slope wetlands. Bidirectional surface water flow, controlled by water-level fluctuations such as seiches in the adjoining lake, may occur. When the size of a lake is so small relative to a fringe wetland, the lake may be incapable of stabilizing the wetland's water table. In this situation, lacustrine fringe wetlands are indistinguishable from depressional wetlands. Water loss from lacustrine fringe wetlands occurs through flow returning to the lake after flooding, saturation surface flow, and evapotranspiration. Organic matter may accumulate in areas protected from shoreline wave erosion.

Estuarine fringe wetlands are influenced by tides along coasts and estuaries. They are replaced by riverine wetlands upstream, where tidal influences diminish and stream flow becomes the dominant water source. Other water sources include groundwater discharges and precipitation. Estuarine fringe wetlands flood frequently and seldom dry for significant periods since tidal sources maintain wet conditions. Water is lost from these wetlands by tidal exchange, saturated overland flow to tidal creek channels, and evapotranspiration. Accumulations of organic matter may occur where wave erosion is minimal.

Exercise #41: Water Sources and Flows

Identify wetlands within an Area of Interest. Locate nearby water bodies and waterways. Inspect the sources of wetland water input and output. Identify internal streams and ponds. Identify topographic characteristics conducive to wetland formation.

Exercise #42: Wetland Class
Using characteristics discussed above, identify the wetland class present.

Climate

Climate must always be considered when evaluating abiotic characteristics important to plant growth. When comparing two areas of interest that are located close together, such as a wetland at the base of a hill and an adjacent upland, their climate characteristics are generally similar. However, variations in microclimate between the two areas may be of considerable importance. Some insights on microclimate may be obtained by examining landform characteristics (see discussions under **Landforms and Microfeatures**).

When comparing two areas that are some distance apart, differences in the regional climate need to be considered. Characteristics, such as the amount and periodicity of precipitation, the amount of snowfall, the maximum and minimum temperatures, and the growing degree days, can be quite different. These statistics and many others are available from the National Weather Service.

Growing Degree Days (GDD) are heuristic expressions of the cumulative amount of heat available for growth over a specific time interval. The National Weather Service glossary (https://www.weather.gov/bgm/helpnowdatac) states:

"Growing Degree Days - (Abbrev. GDD) - A corn growing degree day (GDD) is an index used to express crop maturity. The index is computed by subtracting a base temperature of 50 F from the average of the maximum and minimum temperatures for the day. Minimum temperatures less than 50 F are set to 50, and maximum temperatures greater than 86 F are set to 86. These substitutions indicate that no appreciable growth is detected with temperatures lower than 50 or greater than 86."

Calculations of GDD can be made for species other than corn if the growth temperature range is known. GDD is used here in order to compare two sites of interest. A site with higher monthly GDDs suggests that the vegetation may mature more rapidly than a site with lower monthly GDDs.

Data on many weather parameters are available at: **https://www.weather.gov/wrh/climate**. Click on the appropriate NWS weather forecast office to bring up weather stations near the area of interest. Click on the "view map" button, zoom toward the area of interest, and click on the "show more stations" button. Zoom in or out until a few stations near the area of interest can be seen. Click on an appropriate station (more than one station may be of interest). Choose "monthly summarized data", the desired "year range", and the variable of interest. Clicking on the "view" button will bring up existing data.

Exercise #43: Precipitation

Record the mean annual and monthly precipitation for each station of interest. Also note the maximum and minimum entries.

Record the seasonal and monthly highest snow depths for each station of interest. Also note the maximum and minimum entries.

Exercise #44: Temperature

For "Monthly Highest Maximum Temperature", record the annual and monthly mean, maximum, and minimum temperatures. Record similar data for "Monthly Lowest Minimum Temperature". Finally, record similar data for "Monthly Mean Average Temperature".

Exercise #45: Growing Degree Days

Record the annual and monthly mean, maximum, and minimum GDDs.

Appendix 1: Soil Drainage Classes

Drainage classes describe the rate at which free water (i.e., water that can drain out by gravitational forces) is removed.

The following definitions are taken from the Soil Survey Manual, (Soil Science Division Staff. 2017).

Excessively drained.—Water is removed very rapidly. Internal free water occurrence commonly is very rare or very deep. The soils are commonly coarse textured and have very high saturated hydraulic conductivity or are very shallow.

Somewhat excessively drained.—Water is removed from the soil rapidly. Internal free water occurrence commonly is very rare or very deep. The soils are commonly coarse textured and have high saturated hydraulic conductivity or are very shallow.

Well drained.—Water is removed from the soil readily but not rapidly. Internal free water occurrence commonly is deep or very deep; annual duration is not specified. Water is available to plants throughout most of the growing season in humid regions. Wetness does not inhibit root growth for significant periods during most growing seasons. The soils are mainly free of, or are deep or very deep to, redoximorphic features related to wetness.

Moderately well drained.—Water is removed from the soil somewhat slowly during some periods of the year. Internal free water occurrence is commonly moderately deep and transitory through permanent. The soils are wet for only a short time within the rooting depth during the growing season but long enough that most mesophytic crops are affected. They commonly have a moderately low or lower saturated hydraulic conductivity in a layer within the upper 1 meter, periodically receive high rainfall, or both.

Somewhat poorly drained.—Water is removed slowly so that the soil is wet at a shallow depth for significant periods during the growing season. Internal free water occurrence is commonly shallow to moderately deep and transitory to permanent. Wetness markedly restricts the growth of mesophytic crops, unless artificial drainage is provided. The soils commonly have one or more of the following characteristics: low or very low saturated hydraulic conductivity, a high water table, additional water from seepage, or nearly continuous rainfall.

Poorly drained.—Water is removed so slowly that the soil is wet at shallow depths periodically during the growing season or remains wet for long periods. Internal free water occurrence is shallow or very shallow and common or persistent. Free water is commonly at or near the surface long enough during the growing season that most mesophytic crops cannot be grown unless the soil is artificially drained. However, the soil is not continuously wet directly below plow depth. Free water at shallow depth is common. The water table is commonly the result of low or very low saturated hydraulic conductivity, nearly continuous rainfall, or a combination of these.

Very poorly drained.—Water is removed from the soil so slowly that free water remains at or very near the surface during much of the growing season. Internal free water occurrence is very shallow and persistent or permanent. Unless the soil is artificially drained, most mesophytic crops cannot be grown. The soils are commonly level or depressed and frequently ponded. In areas where rainfall is high or nearly continuous, slope gradients may be greater.

References

Bechtold, W. A. and Charles T. Scott, 2005. The forest inventory and analysis plot design. Gen. Tech. Rep. SRS-80. Asheville, NC: U.S. Department of Agriculture, Forest Service, Southern Research Station, p. 37-52. Accessed 2023: The forest inventory and analysis plot design | US Forest Service Research and Development (usda.gov).

Bonham, C. D. 2013. Measurements for Terrestrial Vegetation, 2nd Ed. Wiley-Blackwell. 246 pp. Accessed 2023: https://vdoc.pub/download/measurements-for-terrestrial-vegetation-65aomld4tbs0.

Braun-Blanquet, J. 1932. Plant sociology. New York, NY: McGraw-Hill. Accessed 2023: Plant sociology; the study of plant communities; Braun-Blanquet, J. (Josias), 1884- : Free Download, Borrow, and Streaming: Internet Archive. Accessed 2023: https://archive.org/.

Brinson, M. M. (1993). "A hydrogeomorphic classification for wetlands," Technical Report WRP-DE-4, U.S. Army Engineer Waterways Experiment Station, Vicksburg, MS. Accessed 2023: https://wetlands.el.erdc.dren.mil/pdfs/wrpde4.pdf

British Columbia Ministry of Forests and Range and British Columbia Ministry of Environment. 2010. Field manual for describing terrestrial ecosystems. -- 2nd ed. BCMFR Research Branch and BCMOE Resource Inventory Branch, Victoria, B.C. (Reprint with updates 2015). Accessed 2023: https://www.for.gov.bc.ca/hfd/pubs/Docs/Lmh/Lmh25_2015.htm.

Bureau of Land Management. 2021. AIM National

Aquatic Monitoring Framework: Field Protocol for Wadeable Lotic Systems. Tech Ref 1735-2, Version 2. U.S. Department of the Interior, Bureau of Land Management, National Operations Center, Denver, CO. Accessed 2023: https://www.blm.gov/documents/national-office/blm-library/technical-reference/aim-national-aquatic-monitoring-0.

Burns, R. M. and B. H. Honkala, (tech. coords.). Silvics of North America 1. Conifers; 2. Hardwoods. Publication: Agriculture Handbook 654, U.S. Dept. of Agriculture, Forest Service. Accessed 2023: https://www.srs.fs.usda.gov/pubs/misc/ag_654/table_of_contents.htm.

Coles-Richie, M.; Manning, M.E.; Tart, D.; DeMeo, T. 2015. Section 2: Existing vegetation classification. In: Nelson, M.L.; Brewer, C.K.; Solem, S.J., eds. 2015. Existing vegetation classification, mapping, and inventory technical guide, Version 2.0. Gen. Tech. Rep. WO–90. Washington, DC: U.S. Department of Agriculture, Forest Service, Ecosystem Management Coordination Staff. Accessed 2023: https://www.fs.usda.gov/emc/rig/protocols/vegclassmapinv.shtml.

Cowardin, L. M., V. Carter, and F. C. Golet, 1979. Classification of Wetlands and Deepwater Habitats of the United States. U.S . Department of the Interior. Fish and Wildlife Service. Accessed 2023: https://pubs.er.usgs.gov/publication/2000109.

Daubenmire, R. 1968. Plant Communities: a Textbook of Plant Synecology. New York: Harper and Row. Accessed 2023: https://archive.org/search?query=Plant+Communities%3A+a+Textbook+of+Plant+Synecology.

DeYoung, J. 2016. Forest Measurements: An Applied Approach, Accessed 2023:

https://openoregon.pressbooks.pub/forestmeasurements/.

Diaz, Sandra & Kattge, Jens & Cornelissen, Johannes & Wright, Ian & Lavorel, Sandra & Dray, Stéphane & Reu, Björn & Kleyer, Michael & Wirth, Christian & Prentice, I. & Garnier, Eric & Bönisch, Gerhard & Westoby, Mark & Poorter, Hendrik & Reich, Peter & Moles, Angela & Dickie, John & Gillison, Andrew & Zanne, Amy & Gorné, Lucas. 2015. The global spectrum of plant form and function. Nature. 529. 10.1038/nature16489. Accessed 2023: https://www.researchgate.net/publication/287984453_The_global_spectrum_of_plant_form_and_function.

Ellis, Beth, D.Daly, L. Hickey, K. Johnson, J. Mitchell, P. Wilf, & S. Wing. 2009. Manual of Leaf Architecture. Cornell University Press. Accessed 2023: https://www.researchgate.net/publication/270216727_Manual_of_Leaf_Architecture.

Evans, D. & C. K. Brewer. 2015. Appendix D. Image Interpretation in "Existing Vegetation Classification, Mapping, and Inventory Technical Guide." M. L. Nelson, C. K. Brewer, and S. J. Solem, Tehnical Editors and Coordinators. USDA Forest Service. Gen. Tech. Report WO-90. Accessed 2023: file:///C:/Users/grsan/Desktop/Existing%20Vegetation%20Classification%20Mapping%20and%20Inventory%20Technical%20Guide,%20Version%202.0%20Appendixes.pdf.

Federal Geographic Data Committee. 2013. Classification of wetlands and deepwater habitats of the United States. FGDC-STD-004-2013. Second Edition. Wetlands Subcommittee, Federal Geographic Data Committee and U.S. Fish and Wildlife Service, Washington, DC. Accessed 2023: https://www.fws.gov/sites/default/files/documents/Clas

sification-of-Wetlands-and-Deepwater-Habitats-of-the-United-States-2013.pdf.

Fenner, Michael. 1998. The phenology of growth and reproduction in plants. Perspectives in Plant Ecology, Evolution and Systematics, Vol. 1/1. Accessed 2023: doi:10.1078/1433-8319-00053 (uv.mx).

Fire Effects Information System, [online]. U.S. Department of Agriculture, Forest Service, Rocky Mountain Research Station, Fire Sciences Laboratory (Producer). Accessed 2023: https://www.feis-crs.org/feis/.

Francis, J. K. (Editor). Wildland Shrubs of the United States and Its Territories: Thamnic Descriptions: Volume 1. Publication: General Technical Report IITF-GTR-26, U.S. Dept. of Agriculture, Forest Service. Accessed 2023: https://www.fs.usda.gov/treesearch/pubs/27005.

Freeman, B. C. & G. A. Beattie. 2008. An Overview of Plant Defenses against Pathogens and Herbivores. The Plant Health Instructor. DOI: 10.1094/PHI-I-2008-0226-01. Accessed 2023: https://www.apsnet.org/edcenter/disimpactmngmnt/topc/Pages/OverviewOfPlantDiseases.aspx.

Gill, J. D. and W. M. Healy (compiled & revised). 1974. Shrubs and Vines for Northeastern Wildlife. USDA Forest Service General Technical Report NE-9. 1974. Accessed 2023: https://www.srs.fs.usda.gov/pubs/3955.

GPS.gov. GPS Accuracy. Accessed 2022: https://www.gps.gov/systems/gps/performance/accuracy/

Herrick, J. E., Justin W. Van Zee, Sarah. E. McCord, Ericha M. Courtright, Jason W. Karl, and Laura M. Burkett. 2017. Monitoring Manual for Grassland, Shrubland, and Savanna Ecosystems. Vol 1: Core Methods. USDA - ARS Jornada Experimental Range, Las Cruces, New Mexico. Accessed 2023: https://www.blm.gov/documents/national-office/blm-library/technical-reference/monitoring-manual-grassland-shrubland-and.

Hunt, Charles B. 1984. Surficial geology. Report – USGS Unnumbered Series, National Atlas of the United States of America. Accessed: 2023. https://doi.org/10.3133/70211058.

Jackson, S. D., D. J. Henson, D. Hilgeman, M. McHugh, and L. Rhodes, 2022. Massachusetts Handbook for Delineation of Bordering Vegetated Wetlands, Second Edition, Massachusetts Department of Environmental Protection, Bureau of Water Resources, Wetlands Program, Boston, Massachusetts. Accessed 2023: https://www.mass.gov/doc/massachusetts-handbook-for-delineation-of-bordering-vegetated-wetlands/download.

Johnson, Chris: Matthew D. Affolter: Paul, Inkenbrandt: Cam Mosher. 2017. An Introduction to Geology. Libretexts. Accessed 2023: https://geo.libretexts.org/Bookshelves/Geology/Book%3A_An_Introduction_to_Geology_(Johnson_Affolter_Inkenbrandt_and_Mosher).

Johnson, M. J., J. A. Bertrand, & M. M. Turcotte. 2016. Precision and accuracy in quantifying herbivory. Ecological Entomology Vol. 41. Accessed 2023: https://static1.squarespace.com/static/54394d81e4b0b8ecdd4e4c65/t/588bba97d1758e8e694a4a3a/1485552280544/16-+Johnson+et+al+Ecol+Ento+2015.pdf.

Jordan, R. A. & J. M. Hartman. 1995. Safe Sites and the Regeneration of Clethra alnifolia L. (Clethraceae) in Wetland forests of Central New Jersey. Am. Midl. Nat. 133:112-123. Accessed 2022: https://www.jstor.org/stable/2426352.

Junikka, L. 1994. Survey of English macroscopic bark terminology. IAWA Journal, Vol. 15 (1), 1994: 3-45. Accessed 2023: https://www.researchgate.net/publication/52001585_Survey_of_English_Macroscopic_Bark_Terminology.

Keddy, Paul A.. Plant Ecology (Origins, Processes, Consequences). Cambridge University Press. Kindle Edition.

Koch, Elisabeth & Bruns, Ekko & Chmielewski, Frank-M & Defila, Claudio & Lipa, Wolfgang & Menzel, Annette. (2007). Guidelines for Plant Phenological Observations. Accessed 2023: https://www.researchgate.net/publication/266211199_Guidelines_for_Plant_Phenological_Observations.

Leverett, B. and D. Bertolette. ?. Measuring Guidlines Handbook. American Forests. 86pp. Accessed 2023: AF-Tree-Measuring-Guidelines_MW.pdf (americanforests.org).

Lichvar, R. W., N. C. Melvin, M. L. Butterwick, & W. N. Kirchner. 2012. National Wetland Plant List Indicator Rating Definitions. US Army Corps of Engineers. Engineer Research and Development Center. Accessed: https://wetland-plants.sec.usace.army.mil/nwpl_static/data/DOC/NWPL/pubs/2012b_Lichvar_et_al.pdf.

Luttmerding, H. A., D. A. Demarchi, E. C. Lea, D. V. Meidinger, T Vold (eds). 1990. Describing ecosystems in the field. MOE Manual 11. (2nd ed.). . B. C. Ministry of Environment with Ministry of Forests. Accessed 2023: https://a100.gov.bc.ca/pub/eirs/lookupDocument.do?

fromStatic=true&repository=BDP&documentId=13616.

Mueller-Dombois, D.; Ellenberg, H. 1974. Aims and methods of vegetation ecology. New York: John Wiley & Sons. 547 p.

Muscolo, A., S. Bagnato, M. Sidari, & R. Mercurio. 2014. A review of the roles of forest canopy gaps. Journal of Forestry Research (2014) 25(4): 725−736. Accessed 2023: https://www.researchgate.net/publication/278070270_A_review_of_the_roles_of_forest_canopy_gaps.

Natural Resources Conservation Service, National Forestry Handbook, title 190, February 2004. Accessed 2023: https://directives.sc.egov.usda.gov/OpenNonWebContent.aspx?content=37005.wba.

Nock, Charles A; Vogt, Richard J; and Beisner, Beatrix E (February 2016) Functional Traits. In: eLS. John Wiley & Sons, Ltd: Chichester. Accessed 2023: https://www.researchgate.net/publication/296484038_Functional_Traits.

Pérez-Harguindeguy, N., S. Díaz, E. Garnier, S. Lavorel, H. Poorter, P. Jaureguiberry, M. S. Bret-Harte, W. K. Cornwell, J. M. Craine, D. E. Gurvich, C. Urcelay ,E. J. Veneklaas, P. B. Reich, L. Poorter, I. J. Wright, P. Ray, L. Enrico, J. G. Pausas, A. C. de Vos, N. Buchmann, G. Funes, F. Quétier, J. G. Hodgson, K. Thompson, H. D. Morgan, H. ter Steege, M. G. A. van der Heijden, L. Sack, B. Blonder, P. Poschlod, M. V. Vaieretti, G. Conti, A. C. Staver, S. Aquino and J. H. C. Cornelissen. 2013. New handbook for standardised measurement of plant functional traits worldwide. Australian Journal of Botany, 61, 167–234. Accessed: 2023. https://www.researchgate.net/publication/259741364_New_handbook_for_standardise_measurement_of_plant_ functional_traits_worldwide.

Perry, Thomas O. 1989. Tree Roots: Facts and Fallacies. Arnoldia 49:4. Accessed 2023: https://fliphtml5.com/bljm/jjdj/basic.

Raunkiaer, C. 1907. The life-forms of plants and their bearing on geography. Translated from Danish and republished in 1934. In The Life Forms of Plants and Statistical Plant Geography. Oxford: Clarendon Press. Accessed: 2023. https://archive.org/details/in.ernet.dli.2015.271790/mode/2up.

Raunkiaer, C. 1908. The Statistics of life-forms as a basis for biological plant geography. Translated from Danish and republished in 1934. In The Life Forms of Plants and Statistical Plant Geography. Oxford: Clarendon Press. Accessed: 2023. https://archive.org/details/in.ernet.dli.2015.271790/mode/2up.

Raunkiaer, C. 1934. The Life Forms of Plants and Statistical Plant Geography: Being the Collected Papers of Raunkiaer. Translated from the Danish, French and German. Preface by A.G. Tansley. Oxford: Clarendon Press. Accessed: 2023. https://archive.org/details/in.ernet.dli.2015.271790/mode/2up.

Reich, P. B., I. J. Wright, J. Cavender-Bares, J. M. Craine, J. Oleksyn, M. Westoby, and M. B. Walters. 2003. The Evolution of Plant Functional Variation: Traits, Spectra, and Strategies. Int. J. Plant Sci. 164(3 Suppl.):S143–S164. Accessed 2023: https://www.researchgate.net/publication/202000375_The_Evolution_of_Plant_Functional_Variation_Traits_Spectra_and_Strategies.

Riebeek, H. (Design by R. Simmon). 2014. Why is that

Forest Red and that Cloud Blue? How to Interpret a False-Color Satellite Image. Accessed: 2023. https://earthobservatory.nasa.gov/features/FalseColor/page1.php.

Romero, C. 2014. Bark: Structure and Functional Ecology in BARK: USE, MANAGEMENT, AND COMMERCE IN AFRICA. New York Botanical Garden Press. Accessed 2023: https://www.jstor.org/stable/43932771.

Russavage, E., J. Thiele, J. Lumbsden-Pinto, K. Schwager, T. Green, & M. Dovciak. 2021. Characterizing Canopy Openness in Open Forests: Spherical Densiometer and Canopy Photography Are Equivalent but Less Sensitive than Direct Measurements of Solar Radiation. Journal of Forestry, 2021, 130–140. Accessed 2023: https://academic.oup.com/jof/article/119/2/130/6053152.

Schrader, Julian, Peijian Shi, Dana L. Royer, Daniel J. Peppe, Rachael V. Gallagher, Yirong Li, Rong Wang, and Ian J. Wright. 2021. Leaf size estimation based on leaf length, width and shape. Annals of Botany 128: 395–406. Accessed 2023: https://academic.oup.com/aob/article/128/4/395/6307867.

Smith, R. D., Ammann, A., Bartoldus, C., and Brinson, M. M. (1995). "An approach for assessing wetland functions using hydrogeomorphic classification, reference wetlands, and functional indices," Technical Report WRP-DE-9, U.S. Army Engineer Waterways Experiment Station. Accessed 2023: https://erdc-library.erdc.dren.mil/jspui/bitstream/11681/6480/1/8672.pdf.

Vicksburg, MS.

Soil Science Division Staff. 2017. Soil Survey Manual.

Handbook 18. . C. Ditzler, K. Scheffe, and H.C. Monger (eds.). USDA Government Printing Office, Washington, D.C.. Accessed 2023: https://www.nrcs.usda.gov/resources/guides-and-instructions/soil-survey-manual.

Soil Survey Staff, Natural Resources Conservation Service, United States Department of Agriculture. Web Soil Survey. Available online. Accessed 2023: https://www.nrcs.usda.gov/wps/portal/nrcs/main/soils/survey/.

Swain, P. C. 2020. Classification of the Natural Communities of Massachusetts. Massachusetts Division of Fisheries and Wildlife, Westborough, MA. Accessed 2023: https://www.mass.gov/doc/classification-of-the-natural-communities-of-massachusetts.

Tiner, R. W. 1999. Wetland indicators: A guide to wetland identification, delineation, classification, and mapping. Boca Raton, FL: Lewis Publishers, CRC Press. Accessed 2023: https://archive.org/details/wetlandindicator0000tine.

Tiner, Ralph W. 2012. Defining hydrophytes for wetland identification and delineation. United States Army Corps of Engineers. Engineer Research and Development Center (U.S.). Cold Regions Research and Engineering Laboratory (U.S.). Accessed 2023: https://usace.contentdm.oclc.org/digital/collection/p266001coll1/id/4612/.

TRY Plant Trait Database. Accessed 2023: https://www.try-db.org/TryWeb/Home.php.

Tsujimoto, M., K. S. Araki, M N. Honjo, M. Yasugi, A. J. Nagano, S. Akama, M. Hatakeyama, R. S.-Inatsugi, J. Sese, K. K. Shimizu and H. Kudoh. 2020. Genet assignment and population structure analysis in a clonal forest-floor herb, Cardamine leucantha, using RAD-seq. AoB Plants 12(1). Accessed 2023: https://academic.oup.com/aobpla/article/12/1/plz080/5681704.

U.S. Army Corps of Engineers 2020. National Wetland Plant List, version 3.5 Accessed 2023: https://cwbi-app.sec.usace.army.mil/nwpl_static/v34/home/home.html.

U.S. Department of Agriculture, Natural Resources Conservation Service. National soil survey handbook, title 430-VI. Accessed 2023: https://directives.sc.egov.usda.gov.

U.S. Department of Agriculture. Title 430 – National Soil Survey Handbook Part 629 – Glossary of Landform and Geologic Terms, Subpart A – Exhibits. Accessed 2023: https://directives.sc.egov.usda.gov/opennonwebcontent.aspx?content=41992.wba.

U.S. Fish and Wildlife Service. 2019. A System for Mapping Riparian Areas. In The Western United States. U.S. Fish and Wildlife Service – Ecological Services. Accessed 2023: https://www.fws.gov/wetlands/documents/A-System-for-Mapping-Riparian-Areas-in-The-Western-United-States-2019.pdfExercise #:~:text=On%20June%2026%2C%202006%2C%20the%20original%20document%20A,current%20mapping%20and%20technological%20standards%20of%20the%20Service.

United States Department of Agriculture, Natural Resources Conservation Service. 2018. Field Indicators of Hydric Soils in the United States, Version 8.2. L.M. Vasilas, G.W. Hurt, and J.F. Berkowitz (eds.). USDA, NRCS, in cooperation with the National Technical Committee for Hydric Soils. Accessed 2023: https://www.researchgate.net/publication/318239569_Field_Indicators_of_Hydric_Soils_in_the_United_States

USDA Natural Resources Conservation Service. 2008. Hydrogeomorphic Wetland Classification System: An Overview and Modification to Better Meet the Needs of the Natural Resources Conservation Service. Technical Note No. 190–8–76. Accessed 2023: http://observatoriaigua.uib.es/repositori/hum_hydrogeomorphic.pdf.

USDA Natural Resources Conservation Service. Soil Quality Indicators, Soil pH. Accessed 2023: https://nrcspad.sc.egov.usda.gov/DistributionCenter/product.aspx?ProductID=396.

USDA Natural Resources Conservation Service. Soil Quality Test Kit Guide. Accessed 2023: https://efotg.sc.egov.usda.gov/references/Public/WI/Soil_Quality_Test_Kit_Guide.pdf.

USDA, NRCS. 2023. The PLANTS Database (http://plants.usda.gov, 08/25/2023). National Plant Data Team, Greensboro, NC USA.

Valladares, F., L. Laanisto, Ü. Niinemets & M. Zavala. 2016. Shedding light on shade: ecological perspectives of understorey plant life. Plant Ecology & Diversity, Vol.9, No. 3, 237-251. Accessed 2023: https://www.tandfonline.com/doi/full/10.1080/17550874.2016.1210262.

Winthers, E.; D. Fallon; J. Haglund; T. DeMeo; G. Nowacki; D, Tart; M. Ferwerda; G. Robertson; A. Gallegos; A. Rorick; D. T. Cleland; W. Robbie. 2005. Terrestrial Ecological Unit Inventory technical guide. Washington, DC: U.S. Departmentm of Agriculture, Forest Service, Washington Office, Ecosystem Management Coordination Staff. 245 p. Terrestrial Ecological Unit Inventory technical guide. Accessed 2023: https://www.fs.usda.gov/research/treesearch/53335.

Wright I. J, N. Dong, V. Maire, C. Prentice, M. Westoby,

S. Díaz, R. V. Gallagher, B. F. Jacobs, R. Kooyman, E. A. Law, M. R. Leishman, Ü. Niinemets, P. B. Reich, L. Sack, R. Villar, H. Wang, P. Wilf. 2017. Global climatic drivers of leaf size. Science 357: 917–921. Accessed 2023: https://www.researchgate.net/publication/320136242_Global_climatic_drivers_of_leaf_size.

ABOUT THE AUTHOR

Gary Sanford spent his youth on a ranch near Sebastopol, California, where agricultural products included eggs, apples, and cherries. After obtaining a Ph.D. in botany from the University of California, Davis campus, he moved to Massachusetts in the early 1970s, and has spent the last 50+ years in New England. Most of this time was spent working as an environmental consultant and botanist. The past few years have been devoted to non-fiction writing. He co-authored "The Ecology of Common Woody Plants of Cape Cod" and "Chester: A Buddy Forever". He authored "Three Cape Cod Botanical Walks In Dennis, MA", "Three Cape Cod Botanical Walks In Yarmouth, MA","Three Cape Cod Botanical Walks In Barnstable, MA", and "Plant Ecology: Interior Cape Cod".

(email: garysanford43@gmail.com)

www.ingramcontent.com/pod-product-compliance
Lightning Source LLC
Chambersburg PA
CBHW070249230526
45470CB00002B/540